More Harm than Good?

Edzard Ernst · Kevin Smith

More Harm than Good?

The Moral Maze of Complementary
and Alternative Medicine

 Springer

Edzard Ernst
University of Exeter
Exeter
UK

Kevin Smith
School of Science, Engineering
 and Technology
Abertay University
Dundee
UK

ISBN 978-3-319-69940-0   ISBN 978-3-319-69941-7 (eBook)
https://doi.org/10.1007/978-3-319-69941-7

Library of Congress Control Number: 2017957706

Printed on acid-free paper

Copernicus Books is a brand of Springer
The registered company is Springer International Publishing AG
The registered company address is: Gewerbestrasse 11, 6330 Cham, Switzerland

To Danielle

Edzard Ernst

To Louise, Primrose and Abigail

Kevin Smith

Preface

There are hundreds of books on complementary and alternative medicine (CAM). They cover all imaginable aspects of the subject, yet hardly any of them discuss the often-serious ethical problems created by the current popularity of CAM.

Why is that?

Most consumers seem to think that ethical issues are academic, bone dry, uninteresting and not relevant to them. We believe that this view is very wrong, so much so that we have written this book which is entirely focused on the ethical problems that arise in CAM.

Ethical issues in medicine affect all of us, and the more actively that individuals engage with ethical discourse and decision-making the better. However, if medical ethics were merely an academic subject, impenetrable to nonspecialists, it would have little practical value. Accordingly, in this book, we will refer to formal academic ethical theory only where necessary; wherever possible, our discussions will be based on straightforward argumentation and will refer only occasionally to theory. But, where appropriate, we will utilise theoretical approaches to help analyse specific ethical issues that arise in CAM. For those unfamiliar with the principles of medical ethics, a more in-depth introduction is provided in the 'Introduction to Medical Ethics' section which follows this foreword.

In all areas of healthcare—and CAM is no exception—consumers are entitled to expect certain basic ethical precepts to be satisfied. These include the following:

- Competence: healthcare practitioners should be sufficiently skilled and knowledgeable such that their clinical practice is effective and their medical advice is valid and up to date.
- Evidence: proffered treatments and diagnostic procedures should be based upon valid knowledge, obtained through robust processes of scientific research.
- Education: programmes of practitioner education and training should ensure that only qualified, competent practitioners are licensed to practice; these programmes should impart the ability to think critically such that evidence can be evaluated in an impartial fashion.

- Autonomy: patients should be at liberty to choose whether to employ a treatment, with full information being provided to explain how the proposed therapy works, along with its risks and benefits. Additionally, if any other effective therapeutic options exist, these should be fairly presented.
- Honesty: CAM professionals should behave truthfully; this includes practitioners, professional bodies, clinics and sellers of CAM therapies.
- Absence of exploitation: patients, clinical trial participants and consumers should be confident that they will not be taken advantage of or abused.

These essential elements of medical ethics are supported by all of the formal ethical approaches and principles outlined in the next section. As we shall explore in this book, these basic ethical requirements are frequently neglected, ignored or wilfully violated in CAM. We feel it is important to disclose these problems and discuss them critically. Only then can we hope to make progress and hope to serve the best interests of patients and consumers.

Exeter, UK   Edzard Ernst
Dundee, UK   Kevin Smith

Acknowledgements

We owe a debt of gratitude to Prof. David Colquhoun for his generous help with the statistical aspects of this book.

Contents

Introduction to Medical Ethics

This introduction is aimed at those readers who are unfamiliar with the principles, frameworks and approaches used in medical ethics. Those already familiar with the basic ethical concepts used in medicine are invited to ignore this section and instead turn directly to Chapter 1.

Medical ethics is a scholarly discipline and like all academic areas contains its own language and theory. A number of formal theoretical frameworks and principles are utilised by medical ethicists in their analyses of ethical problems. However, in this book our ethical points will be based on straightforward argumentation wherever possible. We will refer to formal academic ethical theory only where necessary and always avoid impenetrable or abstruse theory.

However, since ethical frameworks and principles form the substance of professional ethical discourse, it will be useful for the reader to have a broad understanding of these. And, while avoiding undue reliance on formal ethical theory, we will where appropriate utilise theoretical approaches to help analyse specific ethical issues that arise in CAM.

Throughout this book, our ethical considerations of the issues raised by CAM will be based on an ethical approach known as *utilitarianism*. This ethical framework, which seeks to evaluate the consequences of behaviours and decisions in medicine, is explained in some depth below. Utilitarianism is frequently employed in medical ethics; however, it is not the only approach available or defendable. We think it will be valuable for readers to have some knowledge of the other major ethical approaches that are used in medicine, since this will provide context for the utilitarian approach, and help to show how the other major ethical approaches reach broadly the same basic conclusions about the ethics of CAM as arrived at by utilitarian reasoning.

Thus, we set out below the major approaches used in medical ethics. (It is worth noting that these approaches are also employed in ethics more generally, not only in the domain of medicine.)

Right and Wrong in Medicine

Ethical issues arise wherever the potential for harm or exploitation exists. Virtually all public and private organisations subscribe to ethical standards pertinent to their activities, partially impelled by prominent disasters in which ethical failures were implicated, such as the Enron scandal in which executives engaged in duplicitous accounting that wrecked the company, at a cost to shareholders of billions of dollars. Medicine is no exception: the stakes are high—literally involving life and death—and scandals small and large have occurred throughout the many domains of medicine, ranging from hospitals charging for unnecessary tests and treatment, to medical researchers conducting experiments on unwitting patients, to doctors participating in torture.

The idea that ethics is of central importance in medicine can be traced back to the medical revolution that occurred ancient Greece, when the pioneer physicians of the time set out various standards for physician behaviour, most famously represented by the Hippocratic Oath. Since that time, medical ethics has been codified by several bodies worldwide, and much scholarly work continues to be devoted to the field.

In scientific disciplines, facts prevail—even if there is a great deal of debate prior to the establishment of a given scientific theory, and despite the provisional nature of many scientific theories. By contrast, ethics is an area in which absolute truth is fundamentally unattainable. Ethical rules and policies are inevitably open to argument, and there is some truth in the aphorism 'there are as many ethical positions as there are people'.

But it would be quite incorrect to conclude that subjectivity reigns supreme in the domain of ethics and that there are 'no right and wrong answers'. Ethics is important: in the absence of appropriate ethical standards, much harm and exploitation would inevitably ensue. For example, at the end of the Second World War, war crime trials revealed that Nazi doctors had conducted horrific medical experiments on concentration camp inmates and had participated in coercive medical practices including mass sterilization programmes targeting German citizens. It is quite untenable to suggest that the inherent subjectivity of ethics renders these appalling behaviours in any way acceptable or excusable.

Ethical Frameworks and Principles

Ethical considerations of any problem or issue can be divided into two major categories: (a) nonconsequentialist and (b) consequentialist approaches.[1] Nonconsequentialism considers that the action (or even just the motivation behind

---

[1]Ethicists often refer to nonconsequentialist approaches as 'deontology', and consequentialist approaches as 'teleology'.

an action) is the crucial ethical factor. In other words, the action itself is more important than the actual outcome (consequence) of the action. By contrast, consequentialism holds that outcomes (as opposed to actions) should be the crucial determinants of ethical decisions.

Nonconsequentialism

Nonconsequentialist approaches are based on fundamental principles that are deemed to be ethically correct per se. Such principles create duties that agents ought to follow. These duties are frequently expressed as rules, e.g., 'do not lie'; 'do not kill'; 'first, do no harm'; 'prevent disease wherever you can'; and 'apply, for the benefit of the sick, all measures which are required'.[2] According to advocates of nonconsequentialism, to behave ethically, physicians and everyone else involved in healthcare provision—including CAM practitioners—must follow such rules.

Nonconsequentialist principles differ according to various forms of nonconsequentialism, but some principles are almost ubiquitous amongst all types of nonconsequentialism; examples include fairness, justice and respect for autonomy. The latter principle—often referred to simply as 'autonomy'—is of particular importance in medical ethics. This principle holds that a competent, informed person (such as an adult patient) ought to have the freedom to choose whether or not to consent to a proposed course of action (such as a medical procedure, whether conventional or CAM-based). Further, they should be free to choose between alternatives; for example, if the healthcare market offers a range of different treatments for (say) an injured knee, such as surgical, pharmacological, or physiotherapy-based treatments, the paying patient ought to be allowed to freely choose which treatment to purchase—this being so even if they are likely to make an objectively 'wrong' choice by choosing a treatment that is (based on scientific evidence) less effective than others.

Frequently, nonconsequentialist ethical principles are expressed in the language of *rights*. For example, a patient may assert 'I have the right to be told the truth about my prognosis', or 'I have the right to decline to participate in a clinical trial'. At a policy level, nonconsequentialist ethical principles are often enshrined in ethical codes and laws.

Nonconsequentialism has the advantage of setting clear guidance on ethical behaviour, making it simple for agents to know (or believe) that they are behaving ethically. Nonconsequentialist principles establish clear 'lines in the sand' that (if followed) should prevent the most egregious of activities, such as deceiving patients into participating in clinical trials or failing to provide life-saving treatment. Moreover, in nonconsequentialism the interests of a single individual are protected, even when those are at odds with the interests of a larger group.

However, nonconsequentialism has a number of substantial weaknesses. A major concern is the origin and justification of duties. Take, for example, the duty not to kill. This is enshrined in several seminal religious texts (such as the Christian

---

[2]The penultimate and final rules on this list come from a recent version of the Hippocratic Oath, while 'first, do no harm'—although commonly attributed to the Hippocratic Oath—is of uncertain origin.

Bible), but to secular ethicists this provides no justification whatsoever for the 'do not kill' dictum. Something more is needed to support the rule—and indeed ethicists have several reasons to support a rule that prohibits murder, including (inter alia) loss of a happy life, psychological and possibly material harm to the deceased's family and friends and loss to society of a productive life. But these reasons refer to *consequences* and are thus inherently consequentialist in nature.

Moreover, some duties proposed by nonconsequentialists may find little or no deeper support and thus are difficult or impossible to justify as rules for ethical behaviour. A notorious example would be the various historical 'ethical' rules (and indeed laws) that have sought to vilify and proscribe homosexual activity between consenting adults. Another example would be IVF: this assisted reproduction technique has revolutionised the treatment of infertility, and therefore to any reasonable (nonreligious fundamentalist) person is clearly ethically justifiable. However, IVF inevitably entails the death of embryos and therefore conflicts with the 'do not kill' rule.[3]

Even where nonconsequentialist rules and duties can be individually justified, they may conflict with each other. For example, the 'first, do no harm' rule conflicts with the physician's duty to do their best to save the patient's life: life-saving surgery will often harm the patient. (For example, consider an emergency tracheotomy, in which a choking subject's windpipe is rapidly incised to enable breathing: the subject lives, but their neck and windpipe tissues have sustained traumatic harm.) Nonconsequentialist ethical theory provides no easy answers as to how conflicts between its own rules ought to be resolved.

Another major problem with nonconsequentialism is that rules must be followed even if doing so clearly produces more harm (or less good) than doing otherwise. For example, strict adherence to the 'do not kill' rule renders physician-assisted suicide ethically unacceptable—despite the fact that much suffering could be eliminated by providing medical assistance to patients who wish to end their lives. To take another example, the principle of autonomy entails that patients have an absolute right to refuse to be vaccinated—which must be upheld even where such refusals will contribute to maintaining a reservoir of infection and thus lead to an increase in the numbers of infected individuals and deaths.

In view of the above-mentioned problems associated with nonconsequentialism, we shall not base our ethical considerations of CAM on this approach. We consider consequentialism to be of greater value as a tool for ethical analysis, for reasons that follow. Nevertheless, there are some ethical principles that, while formally nonconsequentialist, are so fundamental that most ethicists and indeed much of the population would consider as valid and important and therefore will be used by us as applicable. Truth-telling is one such example: everyone working in the

---

[3]This is not a hypothetical example: when Robert Edwards, the pioneering scientist responsible for the first IVF baby, was in 2010 awarded a Nobel Prize, the Vatican's then spokesman for medical ethics, Monsignor Ignacio Carrasco, objected strongly, stating that the award was 'completely out of order'; that Edwards was responsible for 'large numbers of freezers filled with embryos in the world', of which most will probably 'end up dead'.

healthcare sector—ranging from physicians to therapists to medical salespeople—is under a moral duty to tell the truth. Sadly, as we will discuss in this book, many individuals in the world of CAM fall short of this duty and earn their income on the back of a variety of mistruths, duplicity and outright lies.

Consequentialism

Consequentialist approaches to ethics require agents to act such as to bring about the best consequences. In doing so, consequentialism to a large extent avoids the above pitfalls associated with nonconsequentialism—although it has its own problems, which we shall consider below.

The main consequentialist theory of relevance to medicine is *utilitarianism*. This moral theory is frequently employed (explicitly or implicitly) in medical decision-making. Utilitarian theory is predicated upon the principle that ethical judgements should be based on the consequences arising (or likely to arise) from a course of action, where desirable consequences are those in which positive *utility* is maximised. Utility refers to the balance of positive mental states ('happiness') versus negative mental states ('suffering') in aggregate across all affected individuals.

While it is not a flawless moral doctrine, utilitarianism possesses a number of features which render it viable as a moral theory. Utilitarianism has high degree of internal consistency. It avoids character judgements. It is highly suited to the rational analysis of ethical problems by weighing costs and benefits. Perhaps most importantly, because happiness and suffering (in all their variant forms) are, respectively, highly valued and strongly deprecated by virtually all agents, utilitarianism holds the potential in principle for universal agreement between all people on ethical matters. (Of course, in reality such agreement is an extremely distant prospect.)

Several variants of utilitarianism exist. Classical (or 'mental state') forms of utilitarianism are focused directly on maximising utility. Other variants seek to maximise utility through indirect or proxy means. For example, *preference-satisfaction utilitarianism* holds that the fulfilment of people's preferences—such as patients being able to choose particular types of treatment—should be the primary goal, on the basis that people are happiest when permitted to make their own choices.

Another variant is *rule-utilitarianism*, where ethical behaviour is defined by clear-cut rules. Utilitarian rules are designed such that adherence to them should maximise utility; but individual agents need not directly try to increase utility—indeed they do not have to understand utilitarianism at all, they simply need to follow the rules. Such rules have some of the advantages associated with non-consequentialist rules, such as simplicity and certainty, while retaining their consequentialist (i.e. utility-boosting) nature.

In most medical contexts, the various subtypes of utilitarianism tend to yield very similar answers when applied to practical problems and thus will not be dealt with separately when considering ethical cases in this book.

As with all ethical systems, utilitarianism is associated with certain criticisms and limitations. One central problem is that utility is difficult to quantify. However, suitable psychological research methods such as questionnaire approaches can be used to estimate happiness and its converse in patients. Ethical decisions do sometimes need to be reached on the basis of less than perfect information, and the fact that happiness and misery cannot be precisely quantified does not amount to grounds for a rejection of utilitarianism.

Another issue for utilitarianism is that it implies a separation between agents and actions. In common morality, and according to virtue ethics theory (see below), the character and motivation of the person facing an ethical decision are deemed important. But utilitarianism generates scenarios in which good consequences can flow from the actions of an agent of poor character and be borne by motivations that are clearly negative. For example, a greedy and selfish surgeon may perform life-saving operations, thus boosting utility, merely motivated by monetary gain and caring nothing about the happiness of his patients.

Utilitarians have attempted to deal with this agent-action disparity in various ways, but these will not be considered here. This is because, to a large extent, in the context of medical ethics the agents vs. actions issue is not a major consideration. In medicine, the agents must (or should) always remain at some distance from the actions involved. In medical ethics, the agents are typically members of an ethics committee (within a hospital, university, or government, etc) or are serving a technical role in terms of 'repairing' patients—this includes physicians, nurses, and other healthcare professionals. Unless a conflict of interest is involved, the personal character of any of these individuals is of no practical relevance: ethics committee members or physicians may be philanderers, be ruthless in business transactions, be spiteful, or never donate to charity—such attributes do not matter as long as they perform their roles effectively and, of course, ethically.

Another problem for utilitarians is that utility deliberations can be highly complex, and this detracts from the 'user-friendliness' of the approach. The complexity arises because utilitarian deliberations must consider all consequences of an action, meaning everything the action brings about not only in terms of utility for the patients but also in terms of the happiness or suffering of others including family and medical staff. But such complexity may simply have to be shouldered, given the strengths of utilitarianism as a tool for ethical analysis. Alternatively, a rule-utilitarian approach (as above) can be adopted, which can avoid individual agents having to constantly make utility calculations in every case presented to them. This latter approach generally works best for busy healthcare practitioners, whereas policy decision-making generally requires more complex utility calculations.

The potential conflict between the happiness of particular individuals and the general happiness presents another difficulty for utilitarianism. If the general happiness is what counts, then it seems to follow that the happiness of individuals may be sacrificed for the happiness of a greater number. Utilitarianism might thus seem to favour a number of repugnant actions or situations, such as compulsory sterilisation of individuals at risk of transmitting a serious genetic disorder, permanent quarantine for those carrying a highly infectious and deadly virus, or even the sacrificing of one patient whose organs can be transplanted to save the lives of several other patients. In such cases, the benefits to the majority—prevention of disease and avoidance of attendant suffering—would appear to outweigh the costs borne by the minority. Thus, the charge is that utilitarianism is inherently unfair, unjust and lacking in respect of individual persons.

However, the question needs to be asked: Does utilitarianism actually imply such reprehensible courses of action? We suggest the answer is emphatically no. In practice, running roughshod over individual liberty and life is bound to impose on the public a state of extreme fear and discontent, and hence seriously undermine the functioning of society. Since such a disastrous outcome is clearly one of great disutility, utilitarianism could not support repugnant actions against individuals, even where a *prima facie* case may be made in favour of such actions on the grounds of the general good. Thus, it would be highly simplistic, or a distortion, to suppose that utilitarianism runs contrary to common ethical notions of fairness, justice and respect.

Although utilitarianism does not imply the blatant sacrifice of the few for the good of the many, some utilitarian ethicists—including the present authors—consider that the utilitarian approach ought to be tempered by an additional principle, namely *respect for autonomy*. We considered this principle above, as it is also a nonconsequentialist principle—meaning that regardless of the effects on overall utility, autonomy must be upheld. But the principle of autonomy is not an ad hoc adjustment to utilitarianism, because autonomy is itself derivable from utilitarian considerations: in the long run, it would be detrimental to the general well-being if the autonomy of individuals were to be routinely disregarded, and conversely allowing people to make their own choices—even though some will make errors (including serious ones)—is likely to be maximally conducive to overall utility. Indeed, the utility-maximising nature of autonomy is the basis for the above-mentioned variant of consequentialism known as preference-satisfaction utilitarianism.

Principlism

The inherent subjectivity of ethics presents a problem: even when reflexive and uncritical responses (such as the 'gut reactions' of tabloid readers) are excluded, along with religious outlooks, medical ethicists still do not all agree on which ethical principles are best.

It was in response to this reality that an approach to ethical analysis known as *principlism* was developed. Principlism attempts to factor in *both* nonconsequentialist and consequentialist approaches. The standard version is based on four core principles:

(1) **Respect for autonomy** (a nonconsequentialist principle—but one that utilitarianism also supports).
(2) **Nonmaleficence** ('first, do no harm'—a nonconsequentialist principle).
(3) **Beneficence** (increase overall utility—a consequentialist principle; it is essentially utilitarianism).

(4) **Justice** (fairly distribute benefits, risks and costs—a nonconsequentialist principle).

Principlists apply this approach to an ethical case by examining how each of the four principles (in turn) applies to the issues raised by the case. It is quite common for medical ethics committees to use this method to reach decisions. Principlism can be very useful as a structured 'checklist' method to address ethical problems and therefore finds favour amongst laypeople or non-ethically trained professionals, who predominate on ethics committees.

By contrast, professional medical ethicists are less likely to use principlism and instead will more often use either a nonconsequentialist or (more frequently) a utilitarian approach. Principlism is somewhat limited by the inherently conflicted nature of the principles upon which it is constructed, meaning that ethicists often see it as too much of an intellectual compromise. It is also a rather slavish approach, demanding that the four principles are addressed in each and every ethical case, whether large or small, and regardless of context. This uniformity of approach is rather barren and generally fails to engage with existing arguments. For these reasons, we will not be employing principlism in this book. Nevertheless, a principlist analysis of each of the many CAM scenarios and areas covered in this book is perfectly possible and would lead to very similar conclusions to those we reach by our broadly utilitarian approach.

Virtue Ethics

We suggested above that the character of an agent is not relevant to utilitarian ethics. By contrast, virtue ethics is an approach that emphasises the role of character. This approach, which lies outside of the nonconsequentialism/consequentialism divide, deems an act to be ethically correct if it is the action that a person of good character would do in the same circumstances. Virtues typically include courage, fortitude, fidelity, generosity, magnanimity, prudence and temperance. Various forms of the theory emphasise different virtues, and a central weakness of virtue ethics is that there is no general agreement on what virtues are.

It is uncommon to see contemporary medical ethicists use virtue ethics to analyse issues in medicine. Nevertheless, the fact remains that humans naturally make moral judgements about character and virtue, and most people are not disinterested in the characters of the medical professionals they encounter. Accordingly, in this book we will occasionally refer to virtue ethics. As will become apparent, this is an area in which many proponents, practitioners and sellers of CAM are found badly wanting.

Ethics and CAM

In all areas of healthcare, certain basic ethical precepts ought to be satisfied, and this applies just as much in the realm of CAM as to any other area of medicine. These fundamental ethical requirements may be summarised as follows: CAM practitioners must be competent and adequately educated; their practice must be based upon good evidence; patients must be accorded autonomy; everyone involved in promoting or practicing CAM must behave honestly; and patients and consumers of CAM must never be exploited.

These basic ethical requirements are supported by all of the academic ethical approaches and principles discussed above. Where breached by the nonvirtuous behaviour of those engaged in promoting, practicing or selling CAM, several negative corollaries ensue, as follows: autonomy is reduced; justice is lessened; patients' rights are trammelled; bodily, psychological and financial harm is caused; and utility is reduced.

Chapter 1
Clinical Competence

To what extent can practitioners and promoters of CAM be relied upon to practice or recommend safe and effective treatments? This is the question that we shall consider in this chapter.

The idea that it is morally required for healthcare professionals to practice safe and effective forms of therapy—in other words, to be competent—is self-evidently correct and incontestable. The ethical reasons for practitioner competence include:

a. To avoid patients being harmed by unsafe therapies;
b. To avoid patients being harmed through failing to benefit from the most effective therapies available;
c. To avoid harm to patients through the promotion of a general belief in 'alternatives' to proven forms of medicine.

Thus, avoidance of harm is the main ethical rationale underpinning the presumption that competence is an ethical requirement. To prevent harm, there exists a moral imperative on those practicing or recommending any form of healthcare to ensure that their knowledge is thorough and up-to-date. More fundamentally, this knowledge must be scientifically and logically valid. Moreover, it can never be sufficient for healthcare practitioners to merely act in good faith: regardless of how sincerely a false medical belief is held, the agent who acts—however honestly—on such a belief is liable to become the subject of justifiable moral opprobrium. For example, consider a religiously motivated physician who insists in treating his patients by intercessory prayer. This physician is behaving in an ethically reprehensible manner, and the fact that he truly believes in prayer as the best form of therapy does not justify this practice, nor does it excuse him morally.

Regrettably, the reality is that many CAM proponents allow themselves to be deluded as to the efficacy or safety of their chosen therapy, thus putting at risk the health of those who heed their advice or receive their treatment. Later in this chapter we shall use short stories illustrating such ethically fraught behaviour, and we will back these up with a few examples from real-life. First, however, it will be useful to examine some of the fundamental issues concerning 'competence' in CAM.

© Springer International Publishing AG 2018
E. Ernst and K. Smith, *More Harm than Good?*
https://doi.org/10.1007/978-3-319-69941-7_1

1   What Does 'Competent' Mean?

Most CAM practitioners believe themselves to be competent. This raises some important questions: amongst CAM advocates and practitioners, what is competence taken to mean? What, for instance, is a competent homeopath? This question was recently addressed by an international team of researchers, who investigated homeopathy-educators' views on what a "competent homeopath" might be and what homeopaths might require in their education (Viksveen et al. 2012). This investigation was based on telephone interviews with 17 homeopathy-educators from different schools in 10 European countries. The two main questions asked were "What do you think is necessary in order to educate and train a competent homeopath?" and "How would you define a competent homeopath?"

The study's authors considered that basic resources and processes of study and self-development contribute to the development of a competent homeopath. Such a practitioner is one who possesses certain knowledge and skills, all underpinned by a set of basic attitudes, including 'openness' and a 'patient-centred' outlook. And they concluded that *'this study proposes a substantive theory to answer what homeopathy educators believe a competent homeopath is and what it takes to be educated and trained to become one. The model suggests that certain basic resources and educational and self-developmental processes contribute to developing knowledge and skills necessary to be competent homeopaths. It also pinpoints underlying attitudes needed in the education as well as the clinical practice of competent homeopaths.'*

Such statements are clearly not specific to homeopathy. We could substitute almost any other healthcare profession for "homeopath", and the above conclusions would still be applicable. Any healthcare discipline could hardly fail to agree on the need for *'basic resources and educational and self-developmental processes'* for the development of practitioners, nor fail to require them to develop *'knowledge and skills necessary to be competent'*. Thus, such statements of clinical competence are of little practical value. In order to ascertain what is really meant by competence within a profession, it is necessary to probe further. We need to ask: What is the content of this teaching and learning? Exactly what knowledge and which skills are mandated by a given profession?

The exact curricular content, body of knowledge and set of skills will of course vary enormously between professions. To take some broad examples from mainstream healthcare: physiotherapists should be able to recommend therapeutic exercises, based on their knowledge of the structure of the body's joints; midwives should be able to assess the health of the foetus in the womb, utilising their skills in foetal monitoring technology; and surgeons should be able to use surgical tools to excise and repair damaged tissue, based on their knowledge of anatomy and physiology. Similar-sounding statements can also be applied to CAM: homeopaths should be able to prescribe highly dilute medicines, based on their knowledge of the homeopathic materia medica; acupuncturists should be able to insert needles intended to alleviate certain symptoms, based on their knowledge of the body's meridians and acupoints; and chiropractors should be able to perform spinal manipulations intended to treat back pain and other ailments, utilising their knowledge of anatomy and possibly also of vertebral subluxations.

The problem is that such profession-specific content is very difficult to judge in terms of its validity, particularly for the layperson. How can one be sure that the physiotherapist's recommended stretching exercise has a good chance of improving a frozen shoulder? How can one know whether having acupuncture needles inserted in the thorax will be effective in reducing the frequency of asthma attacks? An exhaustive examination of each and every treatment recommended or practiced by the multitude of healthcare professions would be impractical, both for the layperson and in the context of this book. Instead, we suggest that the most useful approach is to evaluate each claimed therapeutic modality in the context of a key question: 'does it work'. In this respect, there are two core criteria to look for: plausibility and evidence.

2 Competence and Plausibility

Plausibility is a central feature of medical validity. A plausible treatment has a rational basis—in other words, its claimed mode of action does not conflict with established tenets of logic and reason—and is supported by (or at the very least does not clash with) well-established scientific principles and generally accepted medical knowledge. The concept of medical plausibility can be illustrated by comparing modern drugs with CAM medicines. For example, in individuals at risk of developing cardiovascular disease (CVD), drugs known as statins are often prescribed to reduce the likelihood of CVD developing and thus to reduce mortality. This is a plausible form of healthcare, because statin therapy is compatible with the foregoing criteria. Firstly, there is no conflict with rationality: the posited mechanism of action of statin drug molecules is that they inhibit an enzyme involved in cholesterol synthesis, the effect of which is an improvement in blood lipid profile, and a concomitant reduction in the risk of lipid-associated damage to the cardiovascular system. In other words, there is no reliance upon magical phenomena in order to explain the actions of statins. The second feature of statin plausibility is that the biochemical and physiological pathways involved in the development of CVD, upon which statins are intended to act, have been thoroughly investigated by many independent scientists over many years, to the extent that there is no reasonable doubt as to reliability of our broad knowledge of these pathways and their dynamics. Moreover, a large number of modelling and laboratory studies have shown that statin molecules do indeed possess the ability to inhibit the enzyme in question, at least under laboratory conditions.

Contrast this with aconite 30C, a widely available homeopathic preparation. Aconite is a powerful neurotoxin derived from the *Anconitum* plant; the term '30C' refers to this chemical extract having been diluted to 1 part in 1×10^{60}. Countless homeopathic websites recommend this preparation as a treatment in a wide range of apparently unconnected conditions, including psychiatric symptoms such as anxiety and agitation; neurological symptoms including tingling, numbness and cephalalgia; and infectious conditions including influenza and acute fever. As with all homeopathic preparations, the supposed therapeutic actions of aconite 30C are based on two central principles: the 'law of similars' (or 'homeopathic similitude') and the 'law of infinitesimals'. The former principle holds that a substance able to cause a symptom in healthy subjects can also be used to treat that symptom. (In the case of aconite, the effects of poisoning by this agent include various neurological symptoms and a sensation of fever.) The latter principle holds that a therapeutic substance becomes more potent as it is diluted, provided that the process of dilution is accompanied by a special form of vigorous shaking and striking of the preparation ('succussion').

Aconite 30C provides a good example of an implausible therapy. There are several ways in which this preparation—and homeopathy in general—lacks plausibility:

- Law of similars (similitude): the general idea that a substance which causes particular symptoms can be used to treat an illness with similar symptoms appears to hold intuitive appeal for many. Indeed, it predates homeopathy (a 19th century invention) considerably—for example, Hippocrates (c. 460–370 BC) mentions the concept. However, there are simply no logical grounds at all for believing 'like treats like': it is an implausible concept. Furthermore, the principle of homeopathic similitude is simply a category mistake: it cannot be applied to modern medicine. The principle was conceived 200 years ago when it was unknown that disease exists in fundamentally different forms, each caused by a specific malfunction of a tissue or an organ. In light of this knowledge, we now know that the only correct therapy for an illness is according to its very specific cause and pathogenesis: diseases cannot be treated according to a common rule. Similitude is based upon such a rule, and this adds to its implausibility.
- Homeopathic dilution: the extremely high dilution factors inherent in most homeopathic preparations present a fundamental problem, because straightforward mathematics shows that, in most cases, it is statistically unlikely that even a single molecule of the original substance will be present in the final preparation taken by the patient. In the case of aconite 30C, to receive just one molecule, the patient would have to consume at least 3×10^{30} L of the homeopathic solution—this being more than double the volume of water in all of the world's oceans combined!
- Potentisation: homeopathic preparations such as aconite 30C are produced by sequentially diluting the concentrated *Anconitum* extract. At each stage of dilution, the abovementioned 'succussion' of the preparation is employed. The process of succussion varies amongst homeopaths: the inventor of homeopathy, Samuel Hahnemann, recommended that the solutions be rapped against a hard leather book between dilutions (he used his bible). Succussion is believed by proponents to increase the potency of the product (hence the term 'potentisation'). However, the idea that any form of shaking and striking during sequential

dilution can increase a preparation's therapeutic potential runs counter to fundamental science and logic, as does the associated belief that the more diluted is a substance, the more powerful it is ('less is more', as homeopaths frequently put it). Indeed, these notions are so deeply implausible that to accept them would be tantamount to believing in magic.

- 'Memory' effect: faced with the fundamentally implausible idea of a preparation containing no active molecules nevertheless being biologically active, contemporary homeopaths have posited the ad hoc notion that the solvent used for dilution retains a 'molecular memory' of the original substance. But the idea that water (or any other solvent, such as alcohol) can have a 'memory' is unsubstantiated by any laws or mechanisms known to science. Moreover, aconite 30C and many other homeopathic medicines are frequently taken as pills. These sugar pills are prepared by the addition of a drop of the ultra-diluted solution, which then dries out. Thus, for the aconite 30C pills to have any activity, it would be necessary for the 'memory' of the long-since-vanished molecules from the *Anconitum* extract to be transferred from the water onto the sugar pill. Again, neither physics nor chemistry provides any support whatsoever for such a mechanism. Finally, the 'memory' of the plant extract must pass into the patient's body and exert physiological or biochemical effects: and biological science offers no explanations as to how such 'information' could possibly alter bodily functions. Thus, the 'molecular memory' concept stacks implausibility upon implausibility.

In summary, plausibility serves as an essential criterion by which to judge the validity of the knowledgebase upon which a claimed therapeutic modality is based. While statins are clearly based upon plausible ideas, homeopathic preparations such as aconite 30C are not. To the extent that a practitioner uses or recommends implausible therapies, such a person cannot be 'competent'. It follows that such practitioners are guilty of unethical conduct.

Plausibility, however, is only one of two criteria that ought to be used to judge the validity of a body of medical knowledge. After plausibility, *evidence* is the second core criterion by which the underlying principles of a therapeutic modality should be evaluated.

3 Competence and Evidence

To the present authors, the notion that evidence ought to be regarded as a crucial determinant of treatment efficacy (and safety) appears axiomatically correct, and indeed irrefutable. We are not alone: all mainstream healthcare professions subscribe to this concept, which as a broad approach is commonly referred to as 'evidence-based medicine' (EBM). For example, for a conventional pharmaceutical drug to gain a licence, it must prove its worth in carefully controlled clinical trials involving large numbers of participants (volunteers and patients). Many CAM professions also claim that their practices are evidence-based; however, as we shall see later, the standards and forms of evidence deemed acceptable by many CAM professions are often highly questionable. For example, anecdotal accounts from individual patients are frequently deemed to be valid evidence of efficacy for CAM therapies. Yet neither individual nor collective experience amounts to evidence.

In terms of evaluating anything in medicine or healthcare, plausibility is a first requirement; evidence is a second—but equally essential—requirement. We would argue that the gathering of clinical evidence of efficacy is normally only warranted where the plausibility criterion has been satisfied. In particular, clinical trials research ought not to be conducted on fundamentally implausible 'therapies': to conduct such research is arguably unethical, because it entails enlisting patients on pointless and time-wasting experiments (and for some implausible therapies, possibly exposing the patients to direct risks to their health). A further danger of running such trials—especially with the sort of weakly designed and poorly managed trials that sadly are commonplace in CAM—is that occasional 'positive' results may be generated quite by chance, with such spurious 'findings' (which should be designated 'false-positives') seized upon and trumpeted as evidence of efficacy by the therapy's proponents—a topic to which we shall return in the next chapter.

Where a proposed therapy is plausible, evidence of actual efficacy is always required, as a plausible therapy may fail to work in practice. Such failure is generally attributable to the complexity of the human body; for instance, a drug that works well on isolated human cells in the laboratory may be broken down rapidly in the bloodstream, rendering it ineffective in practice. Taking the earlier example of statins, there has been some debate as to the 'real world' clinical effectiveness of these drugs. The current consensus is that statins are indeed effective; but such a view is entirely dependent upon evidence (in this case from epidemiological studies, augmenting controlled clinical trials data). The fundamental point here is that therapeutic effectiveness must always be demonstrated by reliable evidence.

Together, the concepts of plausibility and evidence provide a basis for assessing the content of teaching and learning within a given healthcare profession. In so doing, these concepts allow an assessment to be made of the competence of practitioners. We would argue that a competent healthcare professional should be defined as one who practices or recommends plausible therapies that are supported by robust evidence.

Returning to the findings of the above survey which investigated homeopathy-educators' views on what a "competent homeopath" might be, notions of plausibility and evidence are conspicuous by their absence. This is perhaps unsurprising, given that homeopathy lacks both plausibility and reliable evidence of efficacy.

For any given healthcare modality, it should be possible to critically analyse the fundamental tenets upon which teaching and practice are based, to reach a judgement on that modality's plausibility and evidence-base. Such critical evaluations have been conducted for all major forms of CAM, and are easily accessible in the form of a wide range of books, articles and web pages. In the present book, we will include various stories and published articles that are intended to illustrate typical instances of CAM healthcare professionals acting in accordance with implausible medical principles unsupported by evidence. It seems clear that such professionals cannot be described as 'competent'. Naturally, a lack of practitioner competence is a serious moral problem, and in the following stories the ethically problematic issues are examined. Going beyond individual healthcare professionals, the same considerations apply equally to organisations that offer advice to patients and promote CAM; accordingly, we shall also consider the role of CAM organisations, including professional bodies and charities, in the context of competence.

4 A Homeopath's Advice on the Treatment of Diabetes

4.1 The Story

Fred has suffered from diabetes for as long as he remembers. He is intrigued to hear from a friend that homeopathy might offer some help for his condition and searches the Internet for information. He finds a website where a homeopath advocates using homeopathy to control blood sugar levels in diabetic patients (Das 2014).

Management of blood sugar
The commonly used remedies are Uranium Nitricum, Phosphoric Acid, Syzygium Jambolanum, Cephalandra Indica etc. These are classical Homeopathic remedies. These are used in physiologically active doses such as Mother tincture, $3\times$ etc. depending up on the level of the blood sugar and the requirement of the patient. Several pharmaceutical companies have also brought in propriety medicines with a combination of the few Homeopathic medicines. Biochemic remedies which is a part of Homeopathy advocates Biocombination No. 7 as a specific for Diabetes. Another Biochemic medicine Natrum Phos $3\times$ is widely used with a reasonable success in controlling the blood sugar. Scientific studies on the impact of homeopathic medicines in bringing down blood sugar are limited, but many of the above remedies have some positive effects either as a stand-alone remedy or as an adjunct along with other medications.

Fred has been fed up for a while with self-injecting insulin and he decides to try this approach instead. He consults a local homeopath who confirms the value of the treatment and prescribes him the desired homeopathic remedies.

Three days later, Fred is found by his friend unconscious in his flat. The friend calls an ambulance, and in hospital Fred is successfully treated for his diabetic coma. The medical team explain to him that it was 'touch and go', had the friend found him a little later, they would not have been able to save his life.

4.2   Plausibility and Evidence

The claim expressed on the website is not only that homeopathy can complement diabetes treatment; the claim is clearly that it can also be a 'stand-alone' therapy and therefore a replacement of conventional anti-diabetic treatments. The reasons given above for the implausibility of homeopathic aconite 30C are applicable to homeopathy in general, and therefore apply also to the various homeopathic 'remedies' for diabetes listed in the homeopathic website.

In addition to the lack of plausibility, there is no evidence at all to support the use of any of these preparations. If diabetic patients put the above claims to the test, many would die. Because there are many millions of diabetics worldwide, this claim has the potential to kill substantial numbers of people. In view of these dangers, we consider such incompetent claims to be ethically reprehensible.

5   Incompetent Advice on Vaccination

5.1   The Story

Anna is uncertain whether she ought to have her newborn baby vaccinated. She finds some reports in the press about vaccination highly unsettling and gets more and more anxious about deciding. On the one hand, she wants to do everything possible to protect her child from potentially serious infections; on the other hand, she certainly does not want to inflict harm to her baby.

A friend of hers, Christine, has a boy aged 6 who has developed autism. Christine is sure that this is due to the measles, mumps and rubella (MMR) jabs her son had as a baby. Consequently, Christine is on a mission to convince everyone she knows that MMR immunizations are dangerous. Anna has sought the advice of her GP who insists that there is no link between the MMR or any other vaccination and autism. He claims that there is very substantial and sound research evidence to prove this beyond doubt. But Christine tells Anna that GPs are biased; they only want to vaccinate children because they are getting money for it, she says.

Anna comes from a family that has always relied on the help from chiropractors. She has been treated by chiropractors many times before, not just for back pain but for all sorts of health problems, evidently with good success. So, she decides to consult her chiropractor and asks his advice about vaccination. The chiropractor has studied his subject for many years, displays all sorts of diplomas on his wall, some of which relate specifically to child care, and seems knowledgeable as well as compassionate. Anna implicitly trusts this man.

He listens to Anna's worries patiently and tells her that she is entirely right to be concerned. If it were his child, he says, he would not vaccinate. In fact, none of his three children were vaccinated for anything, and they are all a picture of good health. The evidence, he explains, is quite clear: vaccination is risky! And he warns:

even if the evidence were not that straightforward, when in doubt, don't do it. The medical establishment spares no expenses to cover up the truth, he explains; there's a lot of articles denying a link, but they are all sponsored by the pharmaceutical industry, who are merely defending their financial interests. Vaccinations have become a billion-dollar industry, and the medical profession is complicit in the cover-up.

Anna goes home, discusses the matter with her husband who also likes chiropractic and drug-free healthcare. Eventually, they both decide not to vaccinate their child.

5.2   Plausibility, Evidence and Ethics

Many chiropractors seem to have a very disturbed attitude towards immunization: *anti-vaccination attitudes still abound within the chiropractic profession. Despite a growing body of evidence about the safety and efficacy of vaccination, many chiropractors do not believe in vaccination, will not recommend it to their patients, and place emphasis on risk rather than benefit* (Lawrence 2012).

This kind of anti-vaccination stance is implausible, as it denies thoroughly established scientific principles of immunological function. It also defies the evidence, which is very clear on the balance of efficacy versus risks of vaccination.

To find out where this anti-vaccination ideology comes from, it is helpful to look into the history of chiropractic. D. D. Palmer, the magnetic healer who 'invented' chiropractic some 120 years ago, left no doubt about his profound disgust for immunization: "*It is the very height of absurdity to strive to 'protect' any person from smallpox and other malady by inoculating them with a filthy animal poison... No one will ever pollute the blood of any member of my family unless he cares to walk over my dead body...*" (Palmer 1910).

His son, B. J. Palmer, provided a more detailed explanation for the chiropractic profession's rejection of immunization: "*Chiropractors have found in every disease that is supposed to be contagious, a cause in the spine. In the spinal column we will find a subluxation that corresponds to every type of disease... If we had one hundred cases of small-pox, I can prove to you, in one, you will find a subluxation and you will find the same condition in the other ninety-nine. I adjust one and return his function to normal... There is no contagious disease... There is no infection...The idea of poisoning healthy people with vaccine virus... is irrational. People make a great ado if exposed to a contagious disease, but they submit to being inoculated with rotten pus, which if it takes, is warranted to give them a disease*" (Palmer 1909).

Such sentiments and opinions are still prevalent, even though they are today expressed differently: *The International Chiropractors Association recognizes that the use of vaccines is not without risk. The ICA supports each individual's right to select his or her own healthcare and to be made aware of the possible adverse effects of vaccines upon a human body. In accordance with such principles and based upon the individual's right to freedom of choice, the ICA is opposed to compulsory programs which infringe upon such rights. The International Chiropractors Association is supportive of a conscience clause or waiver in compulsory vaccination laws, providing an elective course of action for all regarding immunization, thereby allowing patients freedom of choice in matters affecting their bodies and health* (ICA 2016).

Of course, not all chiropractors share implausible and non-evidenced opinions about immunization. In fact, the profession is currently divided over the issue. Some chiropractors now realize that immunization has been one of the most successful preventative interventions in the history of medicine. Many others, however, do still adhere to the gospel of the Palmers. Statements like the following abound on the Internet and elsewhere:

Vaccines. What are we taught? That vaccines came on the scene just in time to save civilization from the ravages of infectious diseases. That vaccines are scientifically formulated to confer immunity to certain diseases; that they are safe and effective. That if we stop vaccinating, epidemics will return...And then one day you'll be shocked to discover that ... your "medical" point of view is unscientific, according to many of the world's top researchers and scientists. That many state and national legislatures all over the world are now passing laws to exclude compulsory vaccines....

Our original blood was good enough. What a thing to say about one of the most sublime substances in the universe. Our original professional philosophy was also good enough. What a thing to say about the most evolved healing concept since we crawled out of the ocean. Perhaps we can arrive at a position of profound gratitude if we could finally appreciate the identity, the oneness, the nobility of an uncontaminated unrestricted nervous system and an inviolate bloodstream. In such a place, is not the chiropractic position on vaccines self-evident, crystal clear, and as plain as the sun in the sky? (O'Shea 2012).

Our fictitious mother from the 'story' above, Anna, might also have consulted a homeopath or a naturopath, and the result might well have been the same. A substantial proportion of both professions is known to oppose immunizations. Homeopaths even promote their very own alternative: homeopathic immunizations or homeoprophylaxis, as they call it. Fran Sheffield, a homeopath from NSW, is one of many examples to choose from. She began her homeopathic studies after "seeing the benefits homeopathy brought to her vaccine-injured child", and is a founding member of 'The Do No Harm Initiative Inc.', a lobby group for 'homeopathic immunisation': *Homeoprophylaxis has a remarkable record of safety—vaccines less so. From the homeopath's point of view they are still associated with risks: the dose is too strong, they have toxic additives, and they're given by inappropriate pathways. Homeoprophylaxis has avoided these problems. It's also versatile, inexpensive, quick to produce and easy to distribute... The time has come for all of us—governments and individuals—to take a closer look at homeoprophylaxis and how it relieves the burden of disease...* (Sheffield 2014).

Because there is general agreement that immunizations are amongst the most successful public health measures in the history of medicine, some professional homeopathic organisations have issued politically correct statements about it. But such words merely hide the fact that most homeopaths are against vaccinations. Here are a few examples:

It is well known that measles is an important development milestone in the life and maturing processes in children. Why would anybody want to stop or delay the maturation processes of children and of their immune systems? (Scheibner 2016).

Homoeopathy offers an option for disease prevention and cure. There is scientific evidence in favour of homoeopathy for prevention of diseases (Bhalerao 2013).

Seek out homeopathic, osteopathic, naturopathic, or Chinese medical constitutional treatment to boost your child's immune system and help them be as healthy as they can be (Pfeiffer 2016).

It is possible to prevent post-vaccination damage by giving the homeopathic dilution of the vaccine shortly before and after the vaccination in the C200 dilution (Smits 2016).

There are many recorded cases of people making dramatic recoveries with homeopathic medicines following a bad reaction to a vaccination. Expert advice from a registered homeopath is usually required (New Zealand Council of Homeopaths 2016).

If your children do get sick, use homeopathy to help their immune system get over it. Homeopathy is very effective in epidemics of acute illness. Either see a homeopath, buy a book on homeopathic acute care, or take a class on acute homeopathic prescribing (Pfeiffer 2016).

Another example of a homeopathic anti-vaccinationist is Dr. Oksana Frolov, graduate of Saint Petersburg State Medical University and the Los Angeles School of Homeopathy. On her blog, she provides detailed advice for people who might be uncertain whether to vaccinate their children. The advice she gives is typical of that proffered by homeopaths in general, and so is worth quoting at length: immunisation… *can cause some very serious side effects including permanent brain damage, epilepsy, autism, and mental retardation. With so many vaccinations being required, doctors often have to administer several shots at a time, which can often result in a disaster. Vaccines, along with the elements that are supposed to create the antibodies, also contain mercury, aluminum, formaldehyde, animal tissue, animal blood, human cell from aborted babies, potatoes, yeast, lactose, phenol, antibiotics and unrelated species of germs that inadvertently get into the vaccines. Do you really want all this to be injected into your child just to prevent him or her from having a chicken pox? Vaccines are said to work by stimulating the body to produce antibodies, which are supposed to protect us from an invasion of harmful germs. Childhood diseases, such as measles, mumps, rubella and chicken pox, affect the immune system in a way that makes most people immune to them for the rest of their lives. Vaccinations, on the other hand, create an artificial immunity that wears off and allows the person to catch the disease later in life… Homeopathy has proved to be very effective in treatment of childhood diseases, as well as other infections. From its earliest days, homeopathy has been able to treat epidemic disease, such as cholera, typhus, yellow fever, and diphtheria, with a substantial rate of success, when compared to conventional treatments* (Frolov 2015).

Some might argue that much of the above evidence of the incompetent nature of CAM advocates' recommendations is anecdotal and therefore not reliable. To study the subject more systematically, Ernst and his team conducted a survey in 2003. Specifically, they investigated what advice UK homeopaths, chiropractors and general practitioners give on MMR vaccination via the Internet. Online referral directories listing e-mail addresses of UK homeopaths, chiropractors and general practitioners and private websites were visited. All addresses thus located received a letter from a (fictitious) patient asking for advice about the MMR vaccination. After sending a follow-up letter explaining the nature and aim of this project and offering the option of withdrawal, 26% of all respondents withdrew their answers. Homeopaths yielded a final response rate of 53% ($n = 77$), compared to chiropractors at 32% ($n = 16$). GPs unanimously refused to give advice over the Internet. No homeopath and only one chiropractor advised in favour of the MMR vaccination. Only two homeopaths and three chiropractors even indirectly advised in favour of MMR. Homeopaths were worse than chiropractors, with fewer of the former displaying a positive attitude towards the MMR vaccination (Schmidt and Ernst 2003).

A few years later, a team of Canadian public health experts conducted a similar investigation. They tested whether consultation with a medical professional increases the likelihood of receiving a flu shot among women who have given birth in the past five years and to determine whether this association differs by type of medical professional. Data were obtained from the Canadian Community Health Survey and used to examine the association between receiving a flu shot in the past 12 months and consulting with family doctors, specialists, nurses, chiropractors, or homeopaths/naturopaths. Among the 6925 women included in this research, 1847 (28.4%) reported receiving a flu shot in the past 12 months. The results showed that women who in the previous 12 months had consulted with a family doctor were more likely to have received flu shots compared with women who had consulted with a chiropractor or a homeopath or a naturopath. Consultation with family doctors was found to have the strongest association with annual flu shots among women in contact with young children, whereas consultation with alternative care providers was found to have an inverse association. Given the influenza-associated health risks for young children, the influence of CAM practitioners on immunization choice is ethically concerning (Chambers et al. 2010).

5.3   Ethics and Vaccination

In the context of decisions of vaccination, the ethical principle of individual autonomy is in tension with the principle of maximizing wellbeing. The freedom of adults to decide what is best for themselves will result in reduced wellbeing for some of those who decide to heed to advice of CAM proponents and not be immunized. However, many medical ethicists consider autonomy to be a principle which should not be overridden in the 'interests of the patient'—even where a patient is likely to have a poorer medical outcome as a result of their decision. A countervailing view, held by some medical ethicists, is that it is often ethically correct to resort to coercive paternalism in order to bring about the best medical outcomes for patients (Conly 2013). The justification for this authoritarian approach comprises two main components. The first component is the observation that people frequently make mistakes in their decision-making, due either [a] to innate cognitive error tendencies (of the sort that psychologists have repeatedly demonstrated to be present in all people), or [b] to undue influence from others (peer pressure, media misreporting, societal norms, advertising, etc.). The second component of justification is that allowing people to make mistakes with bad medical consequences does, in fact, reduce their autonomy, thereby implying that unrestrained autonomy can be self-defeating. To take a stark example, if a traveler is permitted to decline immunization against yellow fever, and as a consequence contracts the disease and dies, her ability to exercise autonomy has been abrogated.

Opponents of coercive paternalism fall into two main groups: [a] those who hold autonomy to be a paramount principle, and therefore inviolable; and [b] those who consider wellbeing as a paramount principle but who claim that, overall, full autonomy is more likely to be generative of utility than would coercive paternalism. For those in group [a], autonomy is an ultimate principle, not requiring of further justification; accordingly, the argument from paternalism can have no purchase on those who take such a position. By contrast, those in group [b] would be prepared to accept paternalism if and when it could be shown to bring about increased utility—i.e. it would be acceptable in specific cases—but hold that, in general, permitting individuals to make their own decisions will bring about the greatest degree of utility. This is so for several reasons, including the observation that utility accrues from allowing people to learn from their mistakes, and the reality that oftentimes apparently 'irrational' decisions can benefit the utility of the individual. For example, preventing an Mr. Jones from smoking would undoubtedly improve his health; but he may gain utility benefits from smoking, such as pleasure and anxiety relief—benefits which he values more that his health; accordingly, coercively preventing Mr. Jones from smoking would *reduce* his utility. Of course, in reality many smokers are not like Mr. Jones, in that they wish they could cease smoking, because they consider their health to be of greater value than any benefits from smoking; in such cases—i.e. those involving addiction—arguments from coercive paternalism appear to be on stronger ground.

Ethics is not a science, and individual readers must decide for themselves how they view the autonomy issue. The present authors are with group [b] above: we are prepared to accept paternalism *only* in cases where it can be clearly shown to bring about increased utility. In keeping with the stance of many medical ethicists, we suggest that autonomy should serve as the 'default' assumption, with strong evidence needed to override patient autonomy in specific cases. In the context of vaccination, where patients wish to decline vaccination due to the influence of CAM, a truly paternalistic approach—in other words forcing patients to be vaccinated—would appear to us to be unjustifiable. Where an individual's anti-vaccination decision, influenced by incompetent advice or false claims from CAM proponents, presents a risk of ill health to that person (only), the increased utility which coerced vaccination would bring (via disease reduction) would be outweighed by reduced utility through the unpalatable methods of coercion (for example legal action, fines, or worse) that would inevitably be required to enforce such an illiberal policy.

It is important to emphasize that the foregoing argument in support of patient autonomy provides no excuse whatsoever to CAM proponents who pedal anti-vaccination propaganda with the intent of influencing people away from immunization. Indeed, in the context of a free society, where patients are not 'protected' by coercive paternalism, it is clear that accurate medical advice is of particularly high importance. Accordingly, CAM individuals and organizations who misinform and misadvise are truly deserving of moral condemnation.

The ethical situation with regards to child vaccination is somewhat different to the case of an adult who declines to be vaccinated. A parent who refuses to have their child vaccinated imposes a risk of serious illness on that child—with the latter having no say in the matter. This could reasonably be described as a form of child abuse, just as failing to feed one's child would be so judged. Thus, the parent who declines vaccination for their child is to some degree morally culpable; however, where anti-vaccination individuals or organizations (such some CAM practitioners and bodies described in this chapter) have misadvised such parents, it seems clear that those responsible for such misinformation must shoulder a large part of the moral blame.

A final and important reason to consider the anti-vaccination stance as unethical arises from the phenomenon of 'herd immunity'. Particularly (although not exclusively) in the case of childhood diseases, prevention of the disease requires close to 100% of the at-risk population being vaccinated: when this is achieved, 'herd immunity' applies, meaning that no individuals in the population will develop the disease. Vaccines tend to exhibit varied effectiveness amongst individuals, with a proportion (typically 5–10%) of vaccinated individuals not gaining effective immunity. It is these unfortunate individuals who are at risk of the contracting the disease concerned (despite having received the vaccine), unless the entire population (or close to it) is vaccinated, in which case the disease will be fully suppressed. Thus, parents who refuse vaccination for their children are not only placing their own offspring at risk, they are also risking the health of other children by failing to contribute to herd immunity. In this context, namely the protection of whole groups of children from serious diseases, the many CAM proponents who promulgate anti-vaccination ideas or who advise parents against immunization are once again guilty of ethically reprehensible behavior.

6   Incompetence of CAM Organisations

6.1   The Story

Mary has suffered from depression for many months, and this condition has significantly reduced her quality of life, sometimes to the point where she has lost the will to live. Her doctor has put her on anti-depressants, but they have caused very unpleasant side-effects. Therefore, she is looking for a treatment that is free of adverse effects. A friend tells her that acupuncture might be a good option, and she decides to investigate.

On the Internet, she finds hundreds of websites that claim acupuncture is effective for depression. Most of these articles sound too good to be true. Being naturally a skeptical person, Mary searches for a reliable source, ideally an organization dedicated to disseminate factual and truthful information about acupuncture. This is when she comes across the 'Acupuncture Now Foundation' (ANF).

The ANF was founded in 2014 by 'experts' from around the world who *were concerned about common misunderstandings regarding acupuncture and wanted to help acupuncture reach its full potential... [their] goal is to become recognized as a leader in the collection and dissemination of unbiased and authoritative information about all aspects of the practice of acupuncture* (Acupuncture Now Foundation 2016).

Mary is impressed; this seems exactly what she has been looking for. The ANF has recently published a 'white paper' which promises 'a review of the research'. The review provides her with the information she needs: *In a recent meta-analysis, researchers concluded that the efficacy of acupuncture as a stand-alone therapy was comparable to antidepressants in improving clinical response and alleviating symptom severity of major depressive disorder (MDD). Also, acupuncture was superior to antidepressants and waitlist controls in improving both response and symptom severity of post-traumatic stress disorder (PTSD). The incidence of adverse events with acupuncture was significantly lower than antidepressants* (Bauer and Koppelman 2016).

After reading this, Mary consults a local acupuncturist who confirms that acupuncture works very well for depressed patients. She decides to try it, discontinues her anti-depressants, and has several treatments. At first her mood seems to improve, but after this initial apparent change she falls back into a deep depression, and kills herself with an overdose.

6.2   Plausibility and Evidence

According to the concepts of Traditional Chinese Medicine (TCM), all illness is the result of an imbalance of the 'life forces' yin and yang. One method to re-balance them and thus return patients back to health is the insertion of thin needles in specific points on the body surface, i.e. acupuncture. Because of these TCM concepts, traditional acupuncturists believe that they can treat all sorts of diseases and symptoms from allergies to shingles and from impotence to depression.

The notion of yin and yang has no grounding in science but is based on ancient Taoist philosophy. As already pointed out above, the idea of a panacea, a 'cure all', is implausible; no single therapeutic modality will ever be efficacious for all human ills. While some neurophysiological attempts exist to explain acupuncture's mode of action for pain control, no such mechanisms exist to explain how acupuncture could be an efficacious therapy for a wide range of conditions with totally different aetiologies. Specifically, there is no plausible mechanism for acupuncture to be a cure for depression.

To this lack of plausibility, we must add the lack of good clinical evidence. In the abovementioned 'white paper', the ANF present as evidence an analysis of the effectiveness and safety of acupuncture in depression, published in 2010 (Zhang et al. 2010). But this analysis was wide open to bias. It has been criticised thus: "*the authors' findings did not reflect the evidence presented and limitations in study numbers, sample sizes and study pooling, particularly in some subgroup analyses, suggested that the conclusions are not reliable*" (Centre for Reviews and Dissemination 2016). Moreover, several other, more trustworthy analyses were not positive, and an authoritative review concluded that "*the evidence is inconclusive to allow us to make any recommendations for depression-specific acupuncture*" (Dennis and Dowswell 2013).

The ANF is but one of hundreds of similar organisations in the realm of CAM which present themselves as unbiased and responsible. The sad truth, however, is that they very cleverly use a veneer of respectability to misinform and thus mislead the public. This is ethically unacceptable.

7 Incompetent Charities

7.1 The Story

Charles has been diagnosed with colon cancer. As he never had any symptoms and feels perfectly healthy, he is stunned by this bad news. After having recovered from the initial shock, he considers his options. His oncologist suggests an operation followed by chemotherapy. His work colleague has just gone through a similar experience, and Charles has seen first-hand what a challenging procedure this is. He feels that this approach would change him from a healthy to a sick person. Therefore, he decides to 'shop around'.

Charles has a natural distrust of quacks and wants to make sure he only acts upon good advice. He feels that charities can be trusted and looks for a charity that advises on cancer therapies. He is impressed with the information on the website of a charity called 'YES TO LIFE'.

This charity states on its website: "*We provide support, information and financial assistance to those with cancer seeking to pursue approaches that are currently unavailable on the NHS. We also run a series of educational seminars and workshops which are aimed at the general public who want to know more and practitioners working with people who have cancer.*" (Yes to Life 2016). This is precisely what Charles has been searching for.

The website informs Charles about many alternative therapies and curative or supportive treatments of cancer. Here are a few examples:

CARCTOL

Carctol is a relatively inexpensive product, specifically formulated to assist cells with damaged respiration, it is also a powerful antioxidant that targets free radicals, the cause of much cellular damage. It also acts to detoxify the system.

LAETRILE

Often given intravenously as part of a programme of Metabolic Therapy, Laetrile is a non-toxic extract of apricot kernels. It is converted to cyanide in the body which is claimed to destroy cancer cells.

MISTLETOE

Mistletoe therapy was developed as an adjunct to cancer treatment in Switzerland in 1917–20, in the collaboration between Dr. I Wegman M.D. and Dr. Rudolf Steiner Ph.D. (1861–1925). Mistletoe extracts are typically administered by subcutaneous injection, often over many years. Mistletoe treatment improves quality of life, supports patients during recommended conventional cancer treatments and some studies show survival benefit. It is safe and has no adverse interactions with conventional cancer treatments.

UKRAIN

A type of low toxicity chemotherapy derived from a combination of two known cytotoxic drugs that are of little use individually, as the doses required for effective anticancer action are too high to be tolerated. However, the combination is effective at far lower doses, with few side effects.

Charles is taken by the option of trying UKRAIN to treat his cancer, as it is apparently an 'effective' therapy' with 'few side effects'. Against the advice of his oncologist, he consults a therapist who specialises in this treatment and receives a series of injections. At the conclusion of the treatment he feels fine.

One year later, he develops symptoms indicative of bowel cancer, consults his GP who orders a colonoscopy. It shows advanced colon cancer. Further tests reveal that the cancer has spread to the liver. Charles has several operations and many courses of chemotherapy. Unfortunately, none of these measures are able to cure his advanced malignancy, and within a few months further tumors develop. Charles dies three years later.

7.2   Plausibility and Evidence

All the above-listed treatments are questionable on grounds of plausibility. Moreover, all of them have been assessed scientifically; it turns out that for none is the evidence anywhere near as good as described by YES TO LIFE. Below are the conclusions from careful reviews produced by medical scientists, including one of the present authors, published in the medical literature.

CARCTOL

The claim that Carctol is of any benefit to cancer patients is not supported by scientific evidence (Ernst 2009).

LAETRILE

The claims that laetrile or amygdalin have beneficial effects for cancer patients are not currently supported by sound clinical data. There is a considerable risk of serious adverse effects from cyanide poisoning after laetrile or amygdalin, especially after oral ingestion. The risk-benefit balance of laetrile or amygdalin as a treatment for cancer is therefore unambiguously negative (Milazzo et al. 2011).

MISTLETOE

None of the methodologically stronger trials exhibited efficacy in terms of quality of life, survival or other outcome measures. Rigorous trials of mistletoe extracts fail to demonstrate efficacy of this therapy (Ernst et al. 2003).

UKRAIN

The data from randomised clinical trials suggest Ukrain to have potential as an anticancer drug. However, numerous caveats prevent a positive conclusion, and independent rigorous studies are urgently needed (Ernst and Schmidt 2005).

The contrast between the negative scientific assessments of these alleged anti-cancer agents and the positive accounts proffered by YES TO LIFE could hardly be starker. This is but one example of the gross mismatch between [a] scientific knowledge and [b] claims emanating from pro-CAM sources; regrettably, many more similar instances exist. Indeed, such informational mismatch serves as a signal feature of much contemporary quackery. As medical science progresses, the quantity of health-related information is becoming ever-larger, and laypeople now have hitherto unparalleled access to much of this information through the Internet. However, due to the natural difficulties of discerning reliable knowledge in the face of such informational overload, shortcuts may be taken by those seeking medical information, in the form of believing that apparently authoritative CAM sources provide valid information on medical matters.

The promulgation of medical fantasies—such as mistletoe as a cancer treatment —is a potential source of serious harm, insofar as it can influence patients to forgo the use of scientifically validated therapies. Even where an ineffective treatment is only used temporarily, with the patient subsequently reverting to conventional medicine (typically when it becomes apparent that the 'alternative' turns out to be useless), valuable time may have been lost. In the case of cancer, such wasted time will generally result in poorer or even catastrophic outcomes. Patients missing out on effective forms of medicine by opting for ineffective therapies presents a central ethical problem for CAM. It is impossible to know the extent to which this occurrence inflicts actual harm amongst the populace, but we would argue that it is likely to be substantial.

In a later chapter on 'Truth', we will explore in more detail the issues associated with bad medical information and pro-quackery propaganda. At the present juncture, we will simply note that those—such as YES TO LIFE—who proffer incompetent advice on medical matters are clearly behaving in an ethically reprehensible manner.

8 True Cases

The stories we have told above are fictional, but such events do happen regularly. For various reasons we have chosen to not use true stories. However, to demonstrate that our stories are not far-fetched, here are a few cases as they happened in real-life.

Case 1: Acupuncture for Low Back Pain

The patient in this case was a 59-year-old male, who underwent a course of acupuncture for his chronic low back pain (Bayme et al. 2014). During the therapy, the patient noted swelling at the point of puncture, but his acupuncturist dismissed the concern. The region continued to swell, and three days later his family doctor diagnosed a bacterial skin infection and prescribed oral antibiotics. The following day the patient's condition worsened—he started to suffer from chills and more intense pain, to the extent that he went to the emergency room. An emergency CT scan revealed an abscess within his abdominal wall.

The patient was given intravenous antibiotics and was taken to the operating room, where widespread gangrene was discovered. Surgery was used to remove the damaged tissues—a process that had to be repeated (with serial returns to the operating room) until the wound had recovered sufficiently. The patient received four units of blood and required 13 days of hospitalization. To date, he suffers from a disfiguring wound of his abdominal wall.

It is highly likely that the cause for this grave illness was improper hygiene while treating the patient with acupuncture. In other words, the acupuncturist failed to practice in a competent manner.

Although rare, this kind of tragic consequence of acupuncture has been seen previously by other researchers. Of course, similar incompetence on the part of conventional healthcare professionals can also lead to patient harm. However, as with most forms of CAM, acupuncture presents itself as a 'gentle' and 'safe' alternative to mainstream medicine, and this presentation, in combination with acupuncture's demonstrable lack of effectiveness, makes such cases of patient harm even more ethically reprehensible.

Case 2: Chiropractic Spinal Manipulation for Neck and Back Pain

This case involves a 27-year-old male with chronic neck and back pain (Bayme et al. 2014). He was referred to a chiropractor who treated him using cervical manipulation. Immediately after such a manipulation, the patient felt a severe neck pain; 30 min after manipulation the patient started feeling numbness in his hands and legs. The patient was admitted to an emergency room with symptoms of progressive weakness in his arms and legs. The patient was scanned using MRI which demonstrated a blood clot in his spinal cord in his neck area. The patient underwent immediate surgery, which led to an improvement of his symptoms and eventual remission.

The authors of this case report concluded that *chiropractic procedures can be dangerous when performed by practitioners who might be only partially trained, who might tend to perform an insufficient patient examination before the procedure, and thus endanger their patients.* Although rare, this kind of severe spinal cord damage immediately following chiropractic manipulation has been described in several published reports. And in some cases, the harm is irreversible, as shown in the following case.

Case 3: Chiropractic Spinal Manipulation for Neck Pain and Tingling

A 45-year-old man had a two-week history of right sided neck pain and tenderness, accompanied by tingling in the hand. The internist's neurological examination revealed nothing abnormal, except for a decreased range of motion of the right arm. The internist referred the patient to a chiropractor, who proceeded by manipulating the patient's neck on successive days. By the morning of the third visit to the chiropractor, the patient reported extreme pain and difficulty walking. Nevertheless, and without performing a new neurological examination or ordering an MRI scan, the chiropractor manipulated the patient's neck for a third time.

Thereafter, the patient immediately experienced complete paralysis of his arms and legs (quadriplegia). An MRI scan revealed severe damage to the spinal cord, and an emergency operation was carried out. However, this was unsuccessful and the patient remained quadriplegic. There was no serious doubt that the paralysis was caused by the chiropractic spinal manipulation.

The patient successfully sued both the internist and the chiropractor, and the total amount of damages awarded was $14,596,000, of which the internist's liability was 5% ($759,182) (Epstein and Forte Esq 2013).

In his defence, the internist claimed that there was no known report of permanent quadriplegia resulting from neck manipulation in any medical journal, article or book, or in any literature of any kind or on the Internet. Even the quickest of literature searches discloses this assumption to be wrong. The first such case seems to have been published as early as 1957 (Benassy and Wolinetz 1957). Since then, numerous similar reports have been documented in the medical literature.

The internist furthermore claimed that the risk of this injury must be vanishingly small given the large numbers of manipulations performed annually. But this argument is pure speculation; under-reporting of such cases is huge, and therefore exact incidence figures are anybody's guess.

Incompetence on the part of conventional healthcare professionals can also lead to similar forms of patient harm—for example, a surgeon may make an avoidable error while operating on a patient's spine. However, as with the above example of acupuncture-induced harm, chiropraxis presents itself as a 'safe' alternative. Coupled with serious questions about the supposed clinical benefits of chiropractic manipulation, the claim that chiropractors use 'safer' therapies renders such cases of patient harm all the more ethically fraught.

At this juncture, it is worth considering the role of statutory regulation. In some jurisdictions this applies to chiropractors; for example, in the UK it is illegal to practise as a chiropractor unless registered with the General Chiropractic Council (GCC), the statutory regulator established by Parliament (General Chiropractic Council 2016). A similar situation exists in the United States, where State-by-State regulation applies. On first sight, statutory regulation may appear to be an unalloyed positive: by setting and monitoring mandatory standards of training, practice and conduct, the regulator seeks to protect patient safety. So, in the above case studies of chiropractic manipulation, the spinal cord damage and associated debility might conceivably have been avoided had the chiropractors involved followed the requirements set by their relevant regulator.

However, regulators such as the GCC do not require scientific evidence that the treatments provided by their members are effective. This is perhaps unsurprising: there is very little sound evidence for the effectiveness of chiropractic manipulation, and the evidence that does exist is very weak in comparison to that associated with most conventional forms of healthcare; accordingly, an obligation to require evidence of clinical effectiveness would mean that very few CAM regulators could exist!

While possibly serving to exclude outright charlatans from clinical practice, statutory regulation of CAM has a major downside: when applied to modalities that are inherently ineffective or implausible, it confers an undeserved stamp of respectability and approval. It is no coincidence that many CAM professional organizations actively strive for statutory regulation, such are the benefits likely to accrue from an impression of official endorsement. Yet statutory regulation of many forms of CAM can at best only ensure the competent delivery of therapies that are inherently incompetent. Thus, governmental conferment of statutory regulation to questionable CAM modalities is ethically problematic.

Case 4: Naturopathy for Leukaemia

Japanese doctors reported the case of 2-year-old girl who died of precursor B-cell acute lymphoblastic leukaemia (ALL), the most common cancer in children (Usumoto et al. 2014).

The child had no remarkable medical history. She was transferred to a hospital because of respiratory distress and died four hours after arrival. Two weeks before her death, she had developed a fever of 39 °C, which subsided after the administration of a naturopathic herbal remedy. One week before death, she developed jaundice, and her condition worsened on the day of death.

With modern medical technology, the 5-year survival rate of children with ALL is nearly 90%. However, in this case, the child's parents had opted for naturopathy instead of evidence-based medicine. They had not taken their daughter to a hospital for a medical check-up or immunisation since she was an infant. If the child had received routine medical care, it is unlikely that ALL would have led to her death at age 2.

This would appear to be a case of harm through failure to seek reliable healthcare. Most medical ethicists would view harm caused by omissions as deserving the same amount of moral opprobrium as harm caused by positive actions. Indeed, the authors of this case-report concluded that *the parents should be accused of medical neglect regardless of their motives.*

9 Conclusion

Incompetence in any aspect of healthcare is *prima facie* ethically unacceptable. Most CAM approaches are based on scientifically and/or logically implausible principles, and lack reliable evidence of efficacy. Accordingly, any person or organisation practicing or promoting such baseless medical modalities simply cannot be considered competent—or ethical.

The concept of 'non-maleficence' is a cornerstone of medical ethics. The concept means that it is unethical to act in such a way as to harm the patient (it is commonly phrased as the maxim 'first, do no harm', a version of which is present in the Hippocratic Oath). Clearly, medicine will frequently harm the patient: for example, a surgeon must incise the patient's skin and muscle in order to reach an internal organ, thereby inflicting harm on these tissues. Thus, an absolutist principle of non-maleficence is untenable in medical ethics. A more useful formulation is that the *balance* of benefit versus harm must be favourable. The above stories and real-life examples illustrate what happens when incompetent CAM practitioners and organisations treat patients or give advice: the harm caused by such actions is greater than the benefit to the patient.

Practitioners and organisations of CAM have a major and often neglected problem with competence. We contend that the deeply flawed nature of the fundamental principles of many forms of CAM, together with a dearth of evidence of efficacy, renders most CAM treatments and advice intrinsically incompetent and thus inherently unethical.

References

Acupuncture Now Foundation (2016) About the ANF. https://acupuncturenowfoundation.org/about-us/. Accessed 14 Oct 2016

Bauer M, Koppelman MH (2016) Acupuncture: more than pain management. Acupuncture Now Foundation. https://acupuncturenowfoundation.org/wp-content/uploads/2016/05/ANFwhitepaper2016.pdf. Accessed 16 Sept 2016

Bayme MJ, Geftler A, Netz U et al (2014) The Perils of complementary alternative medicine. Rambam Maimonides Med J 5(3):e0019. doi:10.5041/RMMJ.10153

Benassy J, Wolinetz E (1957) Quadriplegia after chiropractic manipulation. Rev Rhum Mal Osteoartic 24(7–8):555, discussion 556

Bhalerao R (2013) Vaccination and homoeopathy. https://hpathy.com/homeopathy-papers/vaccination-and-homoeopathy/. Accessed 11 Aug 2017

Centre for Reviews and Dissemination (2016) The effectiveness and safety of acupuncture therapy in depressive disorders: systematic review and meta-analysis. Database of abstracts of reviews of effects (DARE): quality-assessed reviews. http://www.ncbi.nlm.nih.gov/pubmedhealth/PMH0028440/. Accessed 17 Sept 2016

Chambers CT, Buxton JA, Koehoorn M (2010) Consultation with health care professionals and influenza immunization among women in contact with young children. Can J Public Health 101(1):15–19

Conly S (2013) Against autonomy: justifying coercive paternalism. Cambridge University Press, Cambridge, UK

Das E (2014) World diabetes day: four ways how homeopathy gets into root to cure diabetes. The Indian Express. http://indianexpress.com/article/lifestyle/health/world-diabetes-day-four-ways-how-homeopathy-gets-into-root-to-cure-diabetes/. Accessed 16 Sept 2016

Dennis CL, Dowswell T (2013) Interventions (other than pharmacological, psychosocial or psychological) for treating antenatal depression. Cochrane Database Syst Rev (7):CD006795. doi(7):CD006795. doi:10.1002/14651858.CD006795.pub3

Epstein NE, Forte Esq CL (2013) Medicolegal corner: quadriplegia following chiropractic manipulation. Surg Neurol Int 4(Suppl 5):S327–S3279. doi:10.4103/2152-7806.112620

Ernst E (2009) Carctol: Profit before Patients? Breast Care (Basel) 4(1):31–33. doi:10.1159/000193025

Ernst E, Schmidt K (2005) Ukrain—a new cancer cure? A systematic review of randomised clinical trials. BMC Cancer 5:69. doi:10.1186/1471-2407-5-69 [pii]

Ernst E, Schmidt K, Steuer-Vogt MK (2003) Mistletoe for cancer? A systematic review of randomised clinical trials. Int J Cancer 107(2):262–267. doi:10.1002/ijc.11386

Frolov O (2015) Homeopathy for better health. http://www.homeopathyforbetterhealth.com/. Accessed 14 Oct 2016

General Chiropractic Council (2016) The UK-wide statutory body for chiropractors. http://www.gcc-uk.org/. Accessed 14 Oct 2016

ICA (2016) New Draconian California mandatory vaccination law now in effect. International Chiropractors Association. http://www.chiropractic.org/mandatory-vaccination-law. Accessed 16 Sept 2016

Lawrence DJ (2012) Anti-vaccination attitudes within the chiropractic profession: implications for public health ethics. Top Integr Health Care 3(4)

Milazzo S, Ernst E, Lejeune S et al (2011) Laetrile treatment for cancer. Cochrane Database Syst Rev doi(11):CD005476. doi:10.1002/14651858.CD005476.pub3

New Zealand Council of Homeopaths (2016) Views on vaccination. http://www.homeopathy.co.nz/homeopathy/views-on-vaccination/. Accessed 14 Oct 2016

O'Shea T (2012) The chiropractic position on vaccines. Planet Chiropractic.com. http://www.planetc1.com/chiropractic-articles/Chiropractic_Position_On_Vaccines.html. Accessed 16 Sept 2016

Palmer BJ (1909) The philosophy of chiropractic. Palmer School of Chiropractic, Davenport, Iowa

Palmer DD (1910) The chiropractor's adjustor. Portland Printing House Company, Portland, Oregon

Pfeiffer E (2016) Vaccinations: making informed decisions. berkeley homeopathy. http://www.berkeleyhomeopathy.com/vaccinations/. Accessed 14 Oct 2016

Scheibner V (2016) Homeopathy & measles (beneficial effects of measles) quotes. http://whale.to/m/measles3.html. Accessed 16 Sept 2016

Schmidt K, Ernst E (2003) MMR vaccination advice over the Internet. Vaccine 21(11–12):1044–1047. doi:10.1016/S0264410X0200628X

Sheffield F (2014) Two Approaches to Prevention and Immunisation. Hpathy Ezine 11(1), 21 Oct 2016

Smits T (2016) Post-vaccination syndrome. http://www.post-vaccination-syndrome.com/3840/pvs.aspx. Accessed 14 Oct 2016

Usumoto Y, Sameshima N, Tsuji A et al (2014) Medical neglect death due to acute lymphoblastic leukaemia: an autopsy case report. Fukuoka Igaku Zasshi 105(12):234–240

Viksveen P, Steinsbekk A, Rise MB (2012) What is a competent homeopath and what do they need in their education? A qualitative study of educators' views. Educ Health (Abingdon) 25 (3):172–179. doi:10.4103/1357-6283.109798

Yes to Life (2016) Choose a therapy. Yes to life; your options for cancer. http://yestolife.org.uk/all_therapies.php. Accessed 17 Sept 2016

Zhang ZJ, Chen HY, Yip KC et al (2010) The effectiveness and safety of acupuncture therapy in depressive disorders: systematic review and meta-analysis. J Affect Disord 124(1–2):9–21. doi:10.1016/j.jad.2009.07.005

Chapter 2
Research Fundamentals

Research is the gathering of data, information and facts for the advancement or completion of knowledge. By adding to the store of human knowledge, scientific research has great intrinsic value. Research also has substantial practical value, in the guise of beneficial technologies flowing from such knowledge.

From an ethical perspective, research can be judged acceptable or unacceptable. Ethically salient features of research include: the topics selected for study; the investigative approaches utilised; the way the research is conducted; and the way the findings are communicated. Ethical judgements of research are reasonable and necessary: researchers must be accountable to the public, not least because their research relies on the co-operation of and support from the populace. Research must be conducted with integrity and transparency. Integrity demands that there is a clear fit between what researchers say they will do and how they conduct the research. Transparency means researchers are open about the nature of their research and communicate it appropriately. Lapses in research ethics have the potential to significantly harm us all.

Those who have engaged in unethical research are frequently said to be guilty of 'research misconduct'. Such misconduct takes many forms and comes in varying degrees of severity. While research of all kinds can fail to meet ethical standards, we suggest that, in CAM, such failings are far more frequent, and usually much more profound, than is the case for research within mainstream scientific and medical fields.

In this chapter, we shall start by offering a brief and necessarily incomplete description of what good research should look like. We will then consider some key concepts concerning the statistical approaches used to analyse research data. Finally we shall look more widely at the landscape of published CAM research.

© Springer International Publishing AG 2018
E. Ernst and K. Smith, *More Harm than Good?*
https://doi.org/10.1007/978-3-319-69941-7_2

1   Good and Bad Research

Scientific research has three main components: formulate a clear research question; gather data to adequately answer the question; and present an answer to the question. In the context of medicine, research questions are often focused on drugs, treatments and procedures, and frequently ask: "does it work?" To generate progress, data must be analysed, interpreted, and presented to a wider audience through publication in an appropriate academic journal. Crucially, for obtaining the necessary data, a proper plan of action—a scientific protocol—must be devised and implemented. The precise design of a study will depend upon several interlinked factors, the foremost of which are: the nature of the research question; the feasibility of the research activity; and the associated ethical considerations. In the following sections, we shall consider various types of research, and discuss their merits and demerits.

1.1   Nonhuman Studies

Some types of investigation do not involve human patients at all. Commonly referred to as 'pre-clinical', this form of research might involve animals, or isolated human (or animal) cells maintained in a laboratory (in vitro). Such research generally pertains to experiments in which a test substance or therapy (for example a prototype drug) is delivered to animals or cells, and the resulting biological changes recorded. Notwithstanding the various ethical issues associated with animal experimentation (which are beyond the scope of this book), the use of animals as test-beds for prototype drugs or treatments is viewed as highly valuable as a means of obtaining valuable information on safety and effectiveness. However, information from animal research can only be indicative for human health; it requires confirmation (or otherwise) by studies involving human subjects.

The use of cells to test therapeutic agents can also be useful and necessary; however, such research is most valuable in terms of answering fundamental questions about biology, as opposed to testing therapies. This is because, removed from their bodily environment, cells frequently may react in very different ways to a test substance than do humans. In the body, administered substances are subject to alteration in a myriad of ways (gastrointestinal breakdown, liver processing, haematological effects, and so on), and isolated cells may respond very differently to cells in situ, since the latter context involves many additional influences (including localised tissue effects, hormonal responses, immunological factors, and so on). In general, establishment of the safety and effectiveness of a prototype drug proceeds from in vitro cell experimentation, then to animal studies, and finally to human trials. In total, this process typically involves many studies conducted by several groups of specialised scientists.

It would be entirely unwarranted to claim that a given treatment—whether a synthetic prototype drug, an herbal extract, a homeopathic preparation, or another therapy—has a medical benefit based only on cell or animal research. Such claims would be liable to cause substantial harm, by misleading patients and practitioners: the human body obviously differs from a dish of cells or an experimental mouse, and the history of medical research contains thousands of examples of promising agents that worked well in cells and animals, yet failed to be effective in humans—or produced unexpected and sometimes dangerous side-effects. One might therefore suppose that this kind of claim would not occur, but sadly, this is not always so. For example, many CAM proponents claim that because certain plant extracts kill cancer cells in vitro, it follows that cancer patients should take these extracts (Ernst 2013b).

1.2 Anecdotal Evidence

One way of attempting to find out whether and to what extent a given treatment is effective is to try out the drug or therapy on an individual patient, and observe what happens to that person. In fact, many CAM practitioners take such observations as being a source of valid evidence of therapeutic effectiveness. Sometimes, these individual 'case studies' are written-up and published in CAM journals as 'research'. Unfortunately, however, anecdotal evidence of this sort serves as a woefully inadequate source of medical knowledge. Indeed, findings that accrue from this approach are liable to grossly misinform clinical practice; in this way, anecdotal evidence can often be worse than no evidence at all. It is worth considering why research based on observations of individual patients is so thoroughly defective as a source of clinical knowledge, because an understanding of the underlying deficiencies can illuminate some of the necessary features that medical research must possess if reliable clinical information is to be produced. The key features that render anecdotal evidence so useless are as follows.

1.2.1 Unrepresentativeness

An individual patient may respond differently to a therapy than some or most other patients would. These differential responses can be due to genetic or other subtle differences between patients. Thus, it is invalid to generalise about the effectiveness (or safety) of a therapy based on a single anecdotal account. However, unrepresentativeness is one of the milder problems with anecdotal evidence. After all, if it was the only issue, then valid findings could be obtained by considering several anecdotal accounts together. In fact, many CAM enthusiasts feel that combining the results of a large number of anecdotal accounts can yield a reliable picture of clinical effectiveness. But they are sorely mistaken, because anecdotal evidence is undermined by several other intrinsic weaknesses, such that no matter how many individual cases are added together, the erroneousness of the findings does not fundamentally change.

1.2.2 Subjective Reporting

When a patient agrees to be the subject of an individual clinical 'experiment', they naturally hope that the trial therapy will have beneficial effects on their illness. The same goes for the researcher running the study. In fact, both patient and experimenter would be unlikely to engage in the study unless they had some optimism that the therapy will work, and this sets up a situation in which all parties are ripe for self-delusion. The danger is one of subjectivity: prompted by positive expectations and wishful thinking, the patient may self-report beneficial changes in symptoms that are quite imaginary, and the researcher may view the patient's condition through rose-tinted spectacles, unconsciously misrepresenting what the patient has reported.

1.2.3 The Placebo Effect

A dummy agent (placebo), such as a sugar pill or saline solution, can exert positive effects on the patient. For example, patients given a placebo generally report greater improvements in their pain symptoms than do patients who are not treated at all. This phenomenon, which is mediated by psychosomatic processes, is known as the 'placebo effect' or 'placebo response'.[1] Since Dr. Henry K. Beecher in 1955 published an influential paper titled 'The Powerful Placebo' (Beecher 1955), it has become received wisdom that the placebo effect is both strong and beneficial. Looking at a number of studies, Beecher found that 35% of patients who had received a placebo responded positively. Beecher's work has subsequently been criticised by several scientists, the upshot being that the extent of the placebo response is less than is commonly presumed. For example, a substantial scientific review looked for the placebo response in published papers covering a wide range of conditions (Hrobjartsson and Gotzsche 2010); the authors reported that, although in certain settings the placebo effect may be able to improve some subjective outcomes (especially pain and nausea):

> We did not find that placebo interventions have important clinical effects in general.

So, the placebo effect tends to be greatly exaggerated in the common imagination. In reality, it is unreliable, short-lived and limited in magnitude. Despite this, some CAM proponents—especially when faced with convincing evidence that their favoured remedy does not work—fall back on the placebo effect as a justification for nevertheless continuing to treat their patients with the ineffective therapy. We shall return to this last refuge of CAM apologists in a future chapter. For now, the important point is that the possibility of a placebo response renders anecdotal evidence invalid: if the patient improves following treatment, it is simply

---

[1]Although frequently referred to in the singular, the 'placebo effect' in fact comprises several distinct components.

impossible to disentangle the role of the placebo effect (if any) from the treatment's specific effects (if any). By contrast, well-designed clinical trials are structured such as to separate placebo effects from real effects, as we shall discuss later in this chapter.

1.2.4   The 'Get-Better-Anyway' Effect

Left untreated, clinical conditions will naturally tend to change in one of three ways:

1. Get worse over time; for example, cancers and neurodegenerative disorders.
2. Get better over time; for example, clinical depression and musculoskeletal pain.
3. Fluctuate over time (becoming better for a time, and then worse, and then better, in a long-term cycle); for example, migraine and allergies.

Where the typical natural history of a disorder is of either of the latter two types, another serious issue for the anecdotal approach arises. Patients seeking help for a medical complaint are much more likely to do so when their symptoms have become particularly troublesome; consequently, for conditions where the symptoms tend to naturally improve through time, or fluctuate, such patients are likely to experience a reduction in their symptoms following the treatment—even if the treatment was completely ineffective!

Moreover, when we take a lot of measurements, there will inevitably be some 'outliers', values that are not close to the mean but considerably below or above it. If we repeat these measurements, chances are that they have drifted towards the mean. This statistical phenomenon is known as 'regression to the mean', and in the present context refers to the tendency of unusually intense symptoms to return towards a more typical level (or to disappear) over time.

Measurements made by McGorry et al. (2000) looking at the episodic nature of chronic and recurrent low back pain provide a good illustration of this 'get-better-anyway' effect. Patients with low back pain recorded their pain on a 10-point scale every day for 5 months (they were allowed to take analgesics ad lib).

The results for four patients are shown in the following figure from the McGorry paper:

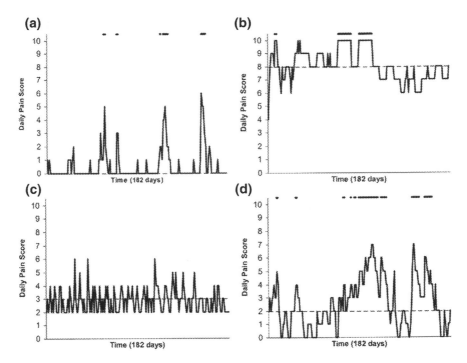

Examples of daily pain scores over a 6-month period for four participants. *Note* Dashes of different lengths at the top of a figure designate an episode and its duration (McGorry et al. 2000)

On average, pain levels stay fairly constant over five months, however they fluctuate greatly, with different patterns for each patient. Painful episodes that last for 2–9 days are interspersed with periods of lower pain or none at all.

Suppose these patients sought treatment from a homeopath at the peak of their pain. After taking the prescribed homeopathic medicine, they would soon feel better —despite the fact that the treatment they received was completely ineffective, and even if it did not produce a detectable placebo response. But the treatment would have been declared a success, even though the patient derived no benefit whatsoever from it. This entirely spurious 'benefit' would be the biggest for the patients whose pain tends to fluctuate the most (i.e. the two patients represented by panels a and d in the above figure).

Thus, the 'get-better-anyway' effect further refutes the notion that worthwhile evidence is to be had from anecdotes. In very many cases, the patient was going to get better anyway; and it is simply not possible, when looking at an individual case study, to disentangle the effects (if any) that a therapeutic intervention has from either the 'get better anyway' effect or the (completely separate) placebo effect.

In terms of actually helping the patient, while the placebo effect can provide some genuine benefits, albeit to a lesser extent than is commonly assumed, the 'get-better-anyway' effect provides *zero benefit* for the patient.

1.2.5 Post hoc ergo propter hoc

Post hoc ergo propter hoc ('after this therefore because of this'), or the 'post hoc fallacy', is a mistake based on the erroneous notion that simply because one event happens after another, the first event was a cause of the second event. It is a form of pattern-recognition that appears to be hardwired into the human brain, as an evolved-in feature that presumably aided our survival as hunter-gatherers by providing a heuristic for rapidly learning about possible causes and effects.[2] While this heuristic helped our prehistoric ancestors, in many domains of modern life it is more of a hindrance than a help. This is certainly the case when it comes to our understanding of cause and effect in the context of health. As the great biologist Peter Medawar said:

> If a person is A) poorly, B) receives treatment intended to make him better, and C) gets better, then no power of reasoning known to medical science can convince him that it may not have been the treatment that restored his health.
>
> Medawar (1967)

In the context of CAM, the *post hoc ergo propter hoc* fallacy interacts with the abovementioned 'get-better-anyway' effect, such that ineffective therapies come to be strongly believed to work. For instance, suppose Mr. Green has been suffering from migraines. Prompted by a spate of particularly bad migraine episodes, he seeks help from a homeopath. He is prescribed homeopathic remedies: because these contain no active ingredients at all, there is nothing in the 'therapy' that can affect Mr. Green's body. Nevertheless, he experiences a marked reduction in migraine symptoms—but this is simply because his condition was poised to improve anyway. No matter: Mr. Green's causality heuristic swings into action, and he comes to strongly believe that homeopathy reduced his migraines. Observing her patient's 'response' to the homeopathy, Mr. Green's homeopath also concludes that homeopathy 'works' for migraine. Suppose the homeopath fancies herself as a researcher; she writes-up Mr. Green's case into an academic paper, which is submitted to a CAM journal which duly publishes the paper. As a result, the entirely unwarranted notion that homeopathy is effective against migraine receives some apparently 'evidence-based' support.

---

[2]For example, suppose an ancestral human eats a toadstool and a few hours later is violently sick. If this experience makes her to believe that the toadstool caused the vomiting, it will aid her survival by helping her avoid a possible source of poisoning in the future. This pattern-recognition mechanism ought to be error-prone in the direction of making false assumptions about causality: if the toadstool is not actually toxic (and the sickness actually resulted from some other unseen cause), the resultant false knowledge carries some cost (in terms of the erroneous avoidance of a potential food source), but this is unlikely to greatly threaten survival. By contrast, if the cause-effect link fails to be made in the case of a genuinely toxic toadstool, the cost is likely be much higher: if more toadstools are eaten, the effects could be lethal. Thus, the genetic sequences underlying the error-prone cause-and-effect heuristic are transmitted to the following generations through natural selection. It is clear that such pressures have influenced the evolution of the human brain such that we have a strong inbuilt tendency to make error-prone assumptions about causality.

The homeopath who treated Mr. Green, or other homeopaths, can collate and publish as many such cases as they can find, but this adds nothing new. Merely cataloguing more and more instances of the same phenomenon—namely apparent clinical effects that may actually be due to the 'get-better-anyway' effect intertwined with *post hoc ergo propter hoc*—should in no way lead to greater confidence in the 'finding' that homeopathy can help improve migraine symptoms.

In summary: attempts to use studies of individual patients as a source of clinical knowledge are dogged by several profound problems, including the potential unrepresentativeness of individual cases, the inevitability of subjective reporting, interference from the placebo effect, improvement in symptoms due simply to the get-better-anyway effect, and erroneous attribution of causality through the strong human tendency to succumb to fallacious *post hoc* reasoning. These elements work in combination to render anecdotal evidence virtually useless, and frequently highly misleading, as a means for evaluating therapeutic effectiveness.

Medical researchers do not usually rely upon isolated studies of individuals. Instead, most studies involve a group of subjects, with data from all of the subjects aggregated in order to show general features. Over the next several sections of this chapter we shall review the major approaches used in medical research to study groups of humans. Some of these methods are stronger than others, as we shall discuss. And the weaker a study is, the more likely it is to run into one or several of the abovementioned pitfalls associated with anecdotal evidence. Robustly designed and rigorously implemented studies are able to transcend the limitations of anecdotal evidence and thus provide reliable medical knowledge; by contrast, the weakest studies—even those that include large numbers of subjects—are no better than an assemblage of anecdotal cases.

Regrettably, in the world of CAM research, the latter type of study predominates. Thus, many published CAM studies contribute only false and misleading information to the body of medical knowledge, as several examples in the next chapter will demonstrate. This is of major ethical concern, because such misinformation leads to aberrant treatment decisions with concomitant harm being inflicted on patients.

1.3 Experimental Versus Observational Studies

For research involving humans, a clinical trial is always the preferable approach for answering research questions about therapeutic effectiveness or safety. Clinical trials are effectively experiments: their aim is to change one variable and observe the consequences. Suppose for example that researchers wish to find out whether a particular agent exerts a therapeutic effect against asthma. This agent could be a prototype pharmaceutical drug or it could be a CAM therapy such as a homeopathic or herbal remedy. In either case, a study could be designed in which a group of asthmatic patients take the agent, and the subsequent number of asthma attacks experienced by these patients is recorded. The precise design of such trials is of crucial importance: we shall discuss the key elements of good clinical trials later in the chapter.

By contrast, 'observational' studies are not experiments, but instead entail the examination of an existing situation. For example, to find out whether omega-3 fatty acids (present in various foods, especially oily fish, and in fish oil supplements) reduce the incidence of asthma attacks, asthmatic patients might be asked to participate in a study in which they are asked to provide information on: [a] their dietary habits and any supplements taken; and [b] how many asthmatic attacks they suffered over a period of time. The researchers would then look at the data for possible correlations between [a] and [b].

1.3.1 Weaknesses of Observational Studies

Observational studies are intrinsically much weaker than their experimental counterparts, for several reasons, the most prominent of which is the non-interventional nature of the study—the researcher is simply looking, as opposed to actively altering variables. This problem means that the data obtained can only show a correlation between a variable (such as the amount of omega-3 consumed) and an observed parameter (such as the frequency of asthmatic attacks). The role of numerous other variables—such as fitness, body mass index, air quality, history of infections, birth difficulties, domestic hygiene, occupation, smoking, alcohol consumption, etc.—cannot be controlled for, and thus a correlation between the variable under observation (i.e. omega-3 intake) and the clinical parameter (asthma attacks) cannot be taken to imply that the first was the cause of the other.

Many observational studies are retrospective in nature: the data depend upon looking backwards in time. This feature promises to inject bias due to the notoriously error-prone nature of memory. Asking individuals to recall past details guarantees interference from well-known cognitive limitations and errors, such as straightforward memory failure, erroneous pattern recognition and wishful thinking. In the omega-3 and asthma example, it is questionable whether trial subjects will be able to remember with sufficient accuracy either details of their diets or how frequent were their asthma attacks over time. This is likely to be compounded by the subjects' awareness of what the study is looking for: the knowledge that there is an 'expected' correlation (e.g. between fish consumption and asthma) is highly likely to unconsciously skew the participants' reports.

So, once again, we run into the same problems that we saw in context of anecdotal evidence (above), in particular the issues associated with subjective reporting and *post hoc ergo propter hoc*. In short, participants may imagine that when they consumed more fish their asthma attacks lessened, and duly tell the researchers so. The resultant 'observational data' obtained from the participants can then be published as seeming to support the notion that omega-3 reduces asthma attacks. But this 'finding' is unwarranted; at best, it may show a very putative link between diet and asthma, which might possibly require further study (using an experimental approach). At worst, such research can pollute the medical literature with misleading associations—which may then be seized upon by those with a vested interest, such as zealous proponents of the supposed health benefits of fish oil and sellers of food supplements.

The confusion of correlation and cause is an elementary and profound logical error. Many things correlate in the absence of a meaningful causal link between them. For example, the incidence of hay fever correlates with sales of ice cream, but it would be wrong to suggest that eating ice cream causes hay fever; rather, it is summer weather that boosts both sales of ice cream and pollen levels leading to hay fever. And many things correlate with no causal link whatsoever.

In most observational studies, there are very many potential factors involved. As outlined above, in the omega-3 and asthma example the researchers will likely have a great deal of data on variables other than omega-3 consumption. Such a wealth of data presents the risk of 'data-dredging': if you look at enough data, you are bound to find correlations—but these correlations will usually be entirely spurious. In statistical terms, data-dredging is a problem of multiple comparisons. In other words, compare enough variables, and you are certain to find some variables that correlate with each other.

To write-up and publish such chance associations as 'real'—i.e. as evidence of cause and effect—is tantamount to spreading medical misinformation. Yet such papers are continuously being published, presenting a general problem for those trying to use research to make sense of factors that influence health. But in the domain of CAM the situation is even worse: the majority of published CAM research studies involving patients (as opposed to papers that are merely based on opinion and speculation) are of the observational type, and most of these are rendered invalid by the sorts of gross weaknesses described above.

1.3.2 Are Observational Studies of Any Value?

So, observational studies are highly limited, and inherently inferior to well-designed experimental studies in terms of being able to disentangle correlation and cause. However, this is not to assert that observational studies are devoid of value. For some research questions, experimental studies are not possible. This is often the case where there is a desire to determine whether health damage is caused by lifestyle or environmental factors. For example, consider the question "do airborne diesel particulates increase the risk of cardiovascular disease?" It would be unethical to investigate this by setting up an experiment deliberately exposing humans to diesel fumes for prolonged periods of time.

Additionally, where a lot of data already exists, it would be unduly stringent to insist that such data are ignored: they might contain some valuable insights, if analysed carefully enough. And some observational studies can be designed such as to minimise the drawbacks described above. For example, if objective clinical information can be obtained—for example by looking at patients' medical records—this can avoid bias associated with participants' memory weaknesses and cognitive errors.

Other positive features of observational studies include that they can often include many more individuals than could a clinical trial, and can cover a much greater timespan. And some observational studies are prospective, rather than retrospective: a cohort of individuals is selected and followed through time, with predefined clinical outcomes assessed at given junctures.

It is hardly surprising that researchers—particularly those with limited resources—frequently resort to an observational approach, where a clinical trial would have been possible. But this is highly regrettable, since observational studies—even where well-designed and executed—can only demonstrate correlation, as opposed to cause and effect. And the problem is particularly rife in CAM research, where claims for the therapeutic effectiveness of any given treatment are typically based on isolated observational studies, which are usually highly flawed, and which never triangulate with additional sources of evidence that might support the notion of a causal link between the therapy and clinical improvement. This will become apparent when we look examples of CAM observational studies in the next chapter.

1.4 Randomised Clinical Trials (RCTs)

Experiments using humans are the best way to obtain reliable information showing cause and effect in medicine. When assessing whether a given treatment works, such experiments are essential. The consensus within medical research is that a particular sort of experimental approach, known as the 'randomised clinical trial' (RCT), is the best tool for this purpose.

In an ideal RCT, patients are randomly assigned to either [a] a group that receives the treatment under investigation or [b] a group—known as the 'control' group—that does not receive the treatment but instead receives a placebo intervention, meaning one that is inert and indistinguishable from the test treatment. The results for group [a] can then be compared with those for group [b], in order to establish whether the treatment worked as hoped for. This means that both groups are exposed to exactly the same range and strengths of influences—except for the specific therapy under investigation.

RCTs can be used to test a substance for its therapeutic effects, such as a newly developed pharmaceutical drug, or a supplement such as omega-3, or a homeopathic remedy: the principles of RCTs are the same, and can be applied to any medicinal therapy. Physical interventions can also be evaluated using RCTs; examples include surgical procedures, or CAM procedures like acupuncture, chiropractic manipulation, or meditation. In such cases, the control group participants might be exposed to a sham version of the treatment under investigation. The choice of the intervention for the control group always depends on the precise research question.

What are the features of a rigorous RCT? Generally speaking, an RCT should be designed such that it allows an unequivocal interpretation of cause and effect.

Clinical study design is a subject in its own right, and an exposition of the detailed aspects and varieties of RCT design is beyond the scope of this book. However, well-founded RCTs have a number of crucial features in common, and we describe these below.

1.4.1   Assignment to Placebo and Experimental Groups

If the research question is 'does therapy X perform better than a placebo', participants should be randomly assigned to two groups (or 'arms'): an experimental arm and a control (or 'placebo') arm. The former subjects receive the test therapy, and the latter subjects receive an indistinguishable and inert version of the treatment. Some RCTs have more than one experimental arm, for example where two drugs are being compared against placebo, or where varying doses of a drug or other treatment are being compared. In all cases, it is essential that adequate numbers of subjects are included in each arm—with the allocations being completely random—in order to maximise the chances of natural biological variation amongst subjects being approximately equalised across the groups.

In clinical research, the use of a control group is crucial: the results from the control arm form the standard against which to compare the results of the experimental arm. Without a valid comparator group, any improvements observed in the experimental group are meaningless because they cannot be attributed to a single cause.

Patients will often show a positive response to a dummy drug or sham procedure. As we discussed above, this effect is most apparent where subjective outcomes are involved. For example, a short-term decrease in patient-reported pain can be reliably obtained by the administration of a placebo. But the effect also occurs with some objective measures; for example, a placebo pill which the subject believes to be an anti-hypertensive drug may cause a short-term decrease in the subject's blood pressure (most likely via psychological anticipation relaxing the subject and thus reducing their blood pressure). This effect is accounted for by the inclusion of a placebo arm in the study.

In addition to genuine placebo effects, a further reason for having a control group stems from the natural trajectory of many disorders and the associated 'get-better-anyway' effect. As discussed above, patients who seek medical attention tend to do so when their symptoms are most troubling. Likewise, it is mostly individuals who exhibit relatively extreme symptoms and signs who are recruited onto clinical trials. This is for two reasons: firstly, these patients, as opposed to their counterparts displaying milder clinical manifestations, are more likely to come into the frame for trial recruitment by dint of their seeking medical attention; secondly, clinical trials normally set thresholds for signs and symptoms, so patients whose conditions (at the time of recruitment) are not severe enough to meet the trial's threshold will be excluded from the study.

Thus, clinical trials tend to be populated with 'extreme' individuals whose symptoms are likely to improve (even if just temporarily) during the clinical trial—even if the experimental therapy is completely ineffective. Note that these improvements are quite separate from any placebo effect; even if none of the trial patients exhibit a placebo response, the 'get-better-anyway' effect entails that improvements will still be observed. Thus, a control group is necessary not only to account for placebo effects, but also to control for the get-better-anyway effect.

The latter phenomenon can be more important than the former in terms of explaining why patients improve following ineffective therapy. Regardless of their relative contribution, it is reasonable to assume that both phenomena work in combination to produce observed improvements in symptoms and signs that have nothing to do with the therapy under investigation. A control group is essential because, on average, the combination of placebo effect plus 'get-better-anyway' effect will be the same in all the groups of a well-designed clinical trial.

Depending on the precise research question, a clinical trial ideally should have a group that receives only a placebo, and no other treatment. However, in some cases, it is unethical to use a pure placebo arm. This applies, for instance, where an illness is serious or life-threatening and for which an effective therapy is available. In such instances, an RCT could assess whether a new therapy has greater effectiveness than the standard treatment. Or all patients may be given standard treatment and one group additionally receives the experimental while the control group gets the placebo in addition.

1.4.2 Double-Blinding

In placebo-controlled trials, subjects must not know to which arm of the RCT they have been allocated, otherwise the whole purpose of having a placebo arm (as above) would be defeated. Patients on the placebo arm must be 'blinded' because the disappointment of not receiving the experimental therapy can lead to negative outcomes or increased dropout rates. The latter would diminish the statistical 'power' of the RCT by reducing the numbers of participants, and therefore reduce the reliability of the trial's findings.

RCTs are said to be 'double-blind' where not only the subjects but also the researchers and all other personnel do not know who is receiving the placebo or the experimental therapy. Double-blinding ensures that patients do not pick up inadvertent cues from physicians or nursing staff which might reveal to the subjects which arm they are on. Also, if the investigators are not blinded, they might consciously or unconsciously treat patients from the two groups differently. For example, positive expectations may be conveyed to subjects on the experimental arm, leading to a skewing of their measured outcomes. Moreover, a lack of double-blinding would increase the risk that unscrupulous researchers will change the results to suit the hoped-for outcome.

Double-blinding is easy to achieve in trials of therapeutic agents such as tablets, liquids or injections by applying randomised codes to the placebos and experimental agents, with the codes only being revealed after the RCT has concluded and the raw data analysed. Ideally, a further safeguard should be implemented: a check made after the treatment to see whether the patients guessed correctly which arm they were on. This would provide information as to how successful the blinding procedures were.

1.4.3 Sufficient Numbers of Participants

RCTs with too few participants—termed 'underpowered' studies—are problematic because they carry a high risk of unreliable results. By analogy, if we ask 80 people to toss a coin once, we can expect a result that is very close to 40 heads and 40 tails. If asking just 8 people to do likewise, there is a much higher chance of a result that diverges from the 'expected' 1:1 ratio. For example, we would certainly not be surprised to obtain a 5:3 ration (5 heads and 3 tails) from such a trial—indeed, this outcome has a 22% probability; whereas the same 5:3 ratio for the 80-person trial (50 heads and 30 tails) has only a 0.7% probability. The analogous risk for underpowered clinical RCTs is that random effects will generate spurious results.

1.4.4 Sufficient Duration

It can take time for medical benefits to develop or side-effects to emerge. However, considering the cost, complexity and inconvenience to subjects, it can be difficult to ensure that RCTs run for sufficiently long periods. Yet RCTs that have overly short treatment periods are of little value. In fact, the results from RCTs of grossly inadequate duration may be of *negative* value, because they frequently amount to nothing more than chance findings which can mislead us all.

Where a therapy is found in RCTs to be effective and thus becomes used in practice, large numbers of patients should be monitored for side-effects and clinical outcome over the following months and years. This allows researchers to detect adverse effects that are too rare to be noticed in relatively small trials of modest duration.

1.4.5 Regulated Environment and Proper Execution

It is important that extraneous factors which could influence the trial results are kept as equal as possible for all the subjects in the RCT. Thus, the ideal study design comprises a highly controlled residential setting, with the patients or volunteers being constantly monitored and provided with near-identical environments.

There is little point in devising a well-designed study if its execution is sloppy or corrupt. The implementation of a study is at least as important as its design. While a poorly devised study should be open to detection by those who read the published report, flawed (or fraudulent) execution is by its nature very difficult to detect. RCTs can be undermined by poor practice at various stages and levels. The most frequent type of major breach of protocol probably involves breakdown of the blinding process, leading to researchers and/or subjects becoming aware of which arm they are in.

Accounting for all the above features can render rigorous RCTs very costly. Many researchers will not have the resources to conduct 'ideal' studies. Lack of resources is particularly common in the CAM research community. Therefore, researchers of CAM frequently cut corners and conduct trials that fall short of the 'ideal'. These suboptimal trials generally comprise very few subjects and have a short duration. Such small-scale trials may, if conforming to the other key requirements of good experimentation (randomisation, double-binding, careful execution, etc.), at best suggest effectiveness. The findings from these mini-RCTs should always be preliminary and contingent upon replication in better, larger studies. But the problem is that such published findings are ripe for cherry-picking, such as to present a misleading or wholly false picture. Unfortunately, this sort of unethical selectivity is commonplace in CAM, as we shall see later.

Despite the consensus amongst medical scientists that the RCT is the 'gold standard' to test effectiveness, other, far less rigorous experimental study designs exist—and are used all too frequently by CAM researchers. For example, a group of patients may be given one type of therapy and the outcome subsequently recorded —but with no control group. This very basic omission means that nobody can know whether the observed outcome was caused by the administered therapy per se or another factor, such as the placebo effect or the 'get-better-anyway' effect.

2 Data Analysis

Normally, researchers must have a protocol that pre-specifies all the methodologies used in a trial, including the planned data analyses. In large-sale medical trials, it is usual to consult a professional medical statistician at the design stage, as well as for the results analysis. But, in CAM, there often is insufficient oversight (or funds) to make sure this always happens. Worse, unscrupulous researchers can frequently find ways to 'torture the data until they confess'. One can run statistical test after statistical test, and eventually one will yield something that can be dressed up as the longed-for positive result. In this section, we shall examine how faulty (or dishonest) reasoning in the context of data interpretation in general and statistical analysis in particular can lead to erroneous and misleading conclusions being reached regarding therapeutic effectiveness.

2.1 Endpoints

It is crucial that, prior to commencement of a study, an appropriate endpoint is defined by the researchers as the one of primary interest. Ideally, such primary endpoints should comprise the most relevant clinical outcomes—in other words, changes that directly improve the clinical condition of the patient. Examples include time to recurrence of cancer, survival time, occurrence of major cardiac event, death rate, or frequency of episodic adverse events (e.g. migraine episode, asthma attacks). More subjective outcomes can also be used, as long as they are measured using validated methods. Examples include pain level, symptom intensity, and depression severity.

Consider, for example, a clinical trial in which herbal remedy X (HRX) is tested as a treatment of angina (the sensation of chest pain, associated with reduced blood flow through the cardiac arteries). A primary endpoint should be established from the outset: for example, the severity of chest pain. Crucially, this endpoint must be expressed in the eventual publication of the study's results.

As is usual in clinical trials, in a study looking at HRX the subjects will also be assessed on several secondary parameters, in addition to the recording of angina pain severity. Some of these measurements will be quite general, and include for example changes in bodyweight, blood pressure, and standard blood parameters, along with subjective scores (e.g. for tiredness and depression). Additionally, more specific supplementary measurements will be ascertained, including an array of physiological and blood biochemistry parameters directly associated angina. For example, levels of the hormone aldosterone may be ascertained, because an increase in aldosterone tends to increase blood pressure, and elevated blood pressure is positively associated with angina. Similarly, sodium levels may be measured, since an increase in blood sodium leads to increased blood pressure.

The assessment of these secondary parameters can be valuable in several ways; for instance helping the researchers check that the experimental therapy is not having any negative impacts on health, and ultimately aiding a more detailed understanding of how the therapy under test actually reduces angina pain—if indeed it proves able to do so.

Unfortunately, the extra data provided by measuring these parameters in addition to the primary endpoint presents a temptation for unethical behaviour. Suppose that HRX fails to reduce angina pain: accordingly, an honest report of the trial would be that no evidence has been produced to show that HRX is effective in reducing angina pain. But suppose that the researchers, wedded to the belief that their therapy is of value, and unwilling to lose hope that they are working on a marketable product, decide to salvage something from the study. Two major options are possible; both of these recourses fall short of outright fraud, but are nevertheless ethically fraught.

Firstly, the researchers may resort to reliance upon the secondary data. While HRX failed to achieve its primary endpoint (reducing angina pain), it might nevertheless have favourably altered another parameter that usually correlates with a reduction in blood pressure and hence with a reduction in angina pain. Suppose a reduction in aldosterone is observed in the experimental group: this seems of some interest, considering the abovementioned link between this hormone and blood pressure. But aldosterone is but one factor in a complex interplay of physiological factors that influence angina pain. The key result from the trial is the finding that HRX fails to reduce angina pain; this result clearly trumps the finding that the HRX is associated with a decrease in aldosterone level.

However, the researchers may be tempted to write-up the trial based on the aldosterone data alone, omitting to mention the primary endpoint. Accordingly, the published report will lend support to the notion that HRX may be of use in combating angina pain. This is likely to lead to unwarranted justifications for more research into a 'promising' potential treatment for angina. Aldosterone level is a 'surrogate endpoint', meaning a measurement that has no guaranteed relationship with the primary endpoint. Selective reporting of surrogate endpoints in this way is clearly misleading, and amounts to unethical behaviour.

The second subterfuge open to the unscrupulous researchers is to scour all the data obtained from the study in an attempt to pick out *any* parameter that shows an apparently 'beneficial' change in patients who received the herbal remedy. This form of research misconduct is data-dredging—a practice we mentioned above in the context of observational studies. It is closely related to the misuse of surrogate endpoints, but it carries a higher risk of generating false medical leads, and thus is even less ethically acceptable. In our example of the HRX trail, perhaps the participants in the experimental arm reported a reduction in depressive symptoms not observed in the control arm. Of course, such a change might be of interest, and could be explored in separate investigations. More likely, however, this change will have come about purely through chance: if one looks at enough random data, as these researchers have done, it will often be possible to find some apparently 'significant' occurrences that are simply statistical artefacts.

The above coin tossing example provides a simple example of this type of problem: if a coin is tossed multiple times, it will often be possible to find an unusually long run of heads. Suppose that six heads in a row occur; the probability of this happening in a single coin tossing trial (of six tosses) is 1.6%, yet in the context of many coin tosses this result is unsurprising and merely a random occurrence. This is analogous to the sifting of a large quantity of data, either from observational or experimental studies, looking for correlations: discovered in this fashion, such correlations are likely to be spurious.

From their unwarranted assumption that HRX reduces depression, our CAM researchers may publish the claim that HRX has potential value as a 'natural' anti-depressant. They will certainly say that their data—from a well-designed RCT no less—argues for more clinical research into a 'promising' potential treatment. And they may even suggest that HRX could be taken by depressed patients; if the researchers don't go this far, we can be sure that other proponents of CAM, including peddlers of herbal remedies, will be recommending that patients take HRX.

But anyone reading the research paper will likely be unaware that the herbal remedy failed to do what it was originally being tested for, i.e. reduce angina pain. Nor will they know that the correlation with depression was merely dredged up from a mass of secondary data, and is therefore most likely to reflect merely a spurious correlation. Note that the researchers are not behaving fraudulently in the sense of having made up their results; but the potential to mislead is very considerable—particularly if the primary endpoint (reducing angina pain) is not made clear in their published report. Clearly, this unethical reporting of a trial is a form of research misconduct.

It is a regrettable fact that the misreporting of secondary data occurs in main-stream medical research. However, such malpractice seems far more common in CAM research. Where lack of therapeutic effectiveness (in terms of the primary endpoint) is endemic, as it often is with CAM therapies, researchers are incentivised to dig into their data in a desperate hunt for something that looks interesting. This problem is exacerbated by the fact that CAM researchers are frequently not edu-cated in, or do not accept, the established conventions of research that deem misuse of surrogate endpoint data and data-dredging as wholly improper.

2.2   P-*Value Pitfalls*

Science as a whole is experiencing something of a crisis: in recent years, it has become clear that a shocking proportion of published claims cannot be replicated by independent scientists.[3] This applies in particular to forms of research that depend fundamentally upon correlations between variables. It was this which prompted John Ioannidis to write a highly influential paper, *Why Most Published Research Findings Are False* (Ioannidis 2005). This strong claim does not apply to all forms of scientific research; studies that are not dependent upon statistical checking of correlations and differences between groups but instead generate categorical results are much less error-prone. For example, the sequencing of a given gene in a blood sample is likely to yield highly reliable data, in the form of a genetic sequence readout; likewise, for the analysis of cellular structure by atomic force microscopy (AFM), in the form of images.

As discussed above, RCTs are the gold standard for investigating therapeutic effectiveness, but RCTs do not produce categorical results; rather, they yield a response difference between the control and experimental groups. The question then remains: is this difference real? In other words, has the experimental treatment has exerted a therapeutic effect? Or is it merely a chance difference between the groups? Differentiating between 'signal' and 'noise' in clinical data depends upon statistical analysis, and wherever this process is flawed, 'false positive' conclusions will be likely. And even when the design and analysis are perfect, the number of false positives will be more than most people expect. In practical terms, this means that ineffective therapies are erroneously reported as being effective.[4]

---

[3]For example, see Nuzzo (2014), and Schwalbe (2016).

[4]There is ongoing academic debate on the extent to which the medical literature is corrupted with false findings. For example, a recent survey of major (mainstream) medical journals claimed that the false positive rate is 14%—a high rate but one that is less than the claim of 'most' published findings being false (Jager and Leek 2014). However, Ioannidis and other academics have repu-diated this claim of 14%, pointing to various flaws in the paper in terms of sampling, calculations, and conclusions, and pointing out that it uses only a very small portion of select papers in top journals (Ioannidis 2014; Benjamini and Hechtlinger 2014).

Faulty statistical interpretation of clinical data serves as an important source of misinformation. A major part of the problem lies in the 'frequentist' statistical methodology commonly used in clinical research. Of particular importance in frequentist statistics is the concept of 'statistical significance'. Tests of significance, such as the t-test, z-test, and chi-squared test, and many others, compute a value known as p *(probability)*. Most papers reporting a clinical trial include a p-value; however, a great deal of misunderstanding exists—even amongst researchers, peer reviewers (academics reviewing submitted papers who are supposed to be experts in the field), and journal editors—over the use and interpretation of p-values. And while this problem afflicts medical research in general, CAM research is particularly prone to major statistical errors, confusions and misrepresentations involving p-values (Pandolfi and Carreras 2014).

In informal terms, a p-value is the probability under a specified statistical model that a difference in clinical response between a treatment group and a control group would be equal to or larger than its observed value would be if the null hypothesis were true. The term 'null hypothesis' refers to the notion that the treatment under test has no clinical effect whatsoever; by contrast, the hypothesis is that the treatment does have a clinical effect.

However, a common misunderstanding of the p-value exists: namely that it is the probability that the null hypothesis is true. Although it is true that the smaller the p-value the less plausible is the null hypothesis, it is also clear that the p-value cannot be the probability that the null hypothesis is true, because it is calculated on the premise that the null hypothesis *is* true. The widespread misunderstanding of what a p-value tells you has led to countless erroneous and misleading claims of therapeutic effectiveness.

The generally accepted (but arbitrary) cut-off threshold is $p = 0.05$; a p-value of above 0.05 is deemed to be merely statistical 'noise', whereas a p-value of 0.05 or lower is considered to be 'statistically significant'—i.e. taken to be evidence of a 'signal' in the data. For example, consider a hypothetical RCT involving a homeopathic remedy for depression: patients in the group receiving the homeopathic remedies (n = 30) show a 19.7% improvement in depression symptoms, while patients receiving placebo pills (n = 30) show only a 15.4% improvement. This result looks quite promising, at least to a homeopath; but a statistical test must be applied to the data. A good test will factor in the number of subjects employed along with the individual responses (not just the average score for each group). Suppose the answer computes as $p = 0.128$. Because this value is greater than 0.05, it is deemed *not* 'significant', and the result of this RCT would therefore be judged as negative: the trial has provided no evidence that the homeopathic remedy improves depression.

The fact that the treatment group showed a numerically greater improvement than the placebo group should not perplex us: it is merely a random occurrence of no significance. Just as tossing a coin 20 times is expected to give 10 heads and 10 tails, skewed results such as 9 heads and 11 tails, or 12 heads and 8 tails, should not surprise us at all—they are merely chance occurrences. If the homeopathy trial had been repeated several times, we would undoubtedly find instances where there was a greater improvement in the placebo group than in the treatment group; in fact, if a very large number of repetitions of this trial were conducted, the number of instances where the placebo gave more positive results versus instances where the homeopathic treatment gave positive results would be expected to be equal.

The critical question about reliance on significance testing is this: what is the *false positive risk* associated with this statistical approach? In other words, we need to know the likelihood of a statistically significant positive trail result being merely a random statistical occurrence.[5] The traditional threshold of $p = 0.05$ is commonly taken to mean that on average 5% (or 1 in 20) of trials will produce a false positive result. This is simply wrong. Of course, given the high stakes involved in clinical research, in terms of erroneously believing that an ineffective treatment is effective, an error rate of 1 in 20 trials would arguably be unacceptably high, considering that thousands of clinical trials are conducted each year.

But this is not the real problem. The notion that $p = 0.05$ equates with a 5% false positive risk is incorrect. For reasons that we shall set out below, even in the case of a perfectly designed and executed RCT, reliance on $p = 0.05$ gives a much higher false positive risk: the odds are that at least 1 in 4 trials will be a false positive. And in many cases the risk will often be much more than this; indeed, in the case of highly implausible CAM therapies, the false positive risk approaches 100%—even where the RCT is perfectly conceived and unbiased in its implementation.[6] As we shall discuss below, two factors are of key importance in explaining why reliance on the *p*-value results in grossly elevated rates of false positive: [a] the 'statistical power' of the trial; and [b] the 'prior probability' of real effects from the therapy under investigation.

2.3   Statistical Power

Statistical power is defined as the probability that a trial will detect a real effect when there is one, i.e. it will give the right result. It can be computed based on the 'effect size'—i.e. the magnitude of change (such as a drug response) that we wish to detect—combined with the sample size—i.e. the number of subjects enrolled on the study. This is one reason why it is important to consult a statistician at the design stage of a trial, and not just at the end; on the basis of effect size, the minimum sample size can be established. The larger the sample size the better, as small samples are more prone to random statistical variation yielding misleading results.

---

[5]In the context of observational studies (as opposed to clinical trials), the equivalent term is *false discovery rate*, in the context of the problem of multiple comparisons.

[6]There is debate amongst statisticians as to whether $p < 0.05$ or $p = 0.05$ is the better interpretation. The former is used by many papers on medical statistics, however it is arguably less realistic than the latter, which tends to generate higher false positive risk values. An in-depth exposition of this subtle but important distinction is beyond the scope of this book; see Colquhoun (2017) for more detailed discussion.

However, especially where expected effect sizes are small, this can lead to the need to recruit very large numbers of participants. This is not just a practical problem, it also has an ethical dimension: the larger the number of participants, the more individuals will be exposed to suboptimal therapy (control group) and side effects (experimental group).

It is no easy task to determine the effect size, as reliable information specific to the therapy under investigation is needed. Ideally, a combination of pre-clinical (in vitro and animal) experimental trials will have been conducted, such as to give a measure of the range of responses to be expected. In some cases, small-scale pilot studies may also have been carried out.[7] Regrettably, in the case of CAM therapies, such groundwork is rarely done. This is partly because CAM research is currently far less active and less well funded than other areas of medical investigation. And one basic factor behind this dearth of preliminary research is the fact that many CAM therapies simply do not actually have specific biological effects, and hence the estimation of 'effect size' becomes nonsensical!

In robust clinical trials, in general a power of 0.8 is aimed for, meaning that a real effect of pre-specified size should be detected in 80% of cases, were multiple tests conducted.[8] In weak trials, of the sort that predominate in CAM research, the power is frequently *much* lower than 80%, due to two factors: [a] an insufficient number of subjects recruited or retained; and [b] a lack of reliable data on biological effects (which may be due to biological effects being nonexistent).

2.4 Prior Probability

In simple terms, prior probability refers to the proportion of tests in which there will be a real effect. It is most easily understood by analogy with diagnostic screening tests (e.g. Colquhoun 2014): the prior probability that has the condition of interest is the fraction of people in the test population that has the condition (i.e. the prevalence of the condition in the population). Homeopathy provides an extreme example of the importance of thinking about the prior probability: because highly diluted homeopathic preparations are identical to placebos, the prior probability of obtaining a real result is zero. By contrast, a pharmaceutical drug that contains highly active ingredients and has proven to be effective in animal studies will have a prior probability substantially greater than zero. A therapy with a prior probability greater than homeopathy but less than the above pharmaceutical drug might be a plant extract reported by an indigenous tribe to have therapeutic effects.

---

[7]It is ethically questionable to conduct such small-scale RCTs, because they are inherently underpowered and thus prone to generating misleading results. Moreover, because effect size and sample size are interrelated, small samples can lead to overestimations of effect sizes. However, in some specific cases the determination of effect size can be aided by data from such trials.

[8]This will only be true if the power calculation has been valid.

The difficulty is that quantification of such probabilities is inherently subjective, and may amount merely to informed guesswork.

As an alternative to reliance upon the fraught p-value, some statisticians have recommended instead utilising Bayesian statistics. The Bayesian approach starts with a quantification of prior probability, which in principle makes this methodology far more robust than a frequentist one. However, the difficulties of accurately estimating a prior probability for a given experimental therapy render this method somewhat impractical in the context of clinical trials. So, even in well-designed clinical studies, a frequentist statistical approach is usually employed. Nevertheless, prior probabilities, although difficult to quantify, do exist in reality.

In general, the higher are [a] the power and [b] the prior probability in a clinical trial, the lower the false positive risk will be. By contrast, carrying out a clinical study into something with a very small prior probability of effectiveness will have a concomitantly high false positive risk.

The magnitude of this problem is great. For example, suppose a researcher estimates (optimistically[9]) that acupuncture has a prior probability of 1% in respect of a specific effect on migraine severity, on the basis that needling may cause a clinically valuable release of endogenous opioids. Further, imagine the researcher conducts a trail of greater rigor than is typical in CAM, where the power of the study is a respectable 80%. The expected false positive risk associated with this trial will be greater than 85%.[10] If the trial is underpowered the risk will be greater still. And for a CAM that is completely implausible—such as homeopathy—the false positive risk approaches or equals 100%.

---

[9]This is optimistic because, as discussed elsewhere in this book, the alleged specific effects of acupuncture are supposedly due to completely implausible physiological features and mechanisms, including Qi ('vital energy') flowing through meridians (body channels), none of which have been discovered by science and all of which are implausible.

[10]Various ways exist to calculate false positive risk; the values that result vary according to methodology, but all valid approaches yield risks that are substantially greater than the 5% assumed by the common but disastrously wrong assumption that $p = 0.05$ equates with a 5% false positive risk. For example, in simplified terms, we can compute the expected false positive risk for $p < 0.05$ by: [a] multiplying the sample size by the prior probability, then multiplying the proportion of the sample with a real effect by the power value, to establish the number of true responders; then [b] multiplying the proportion of the sample who are expected to have showed no effect with the threshold p-value (0.05) to establish the expected number of false positives; and finally [c] expressing as a percentage the number of false positives from the overall number of positive results. For the example given above, this computes to 86%. Note that for $p = 0.05$, using the methodology used by Colquhoun (2017), the false positive risk computes to be even higher, at 97%.

An exposition of the underlying statistical theory is beyond the scope of this book. However, David Colquhoun has examined in persuasive detail the pitfalls of the p-value, and he concludes as follows:

> ...if we observe a P value just below 0.05, then there is a chance of at least 26% that your result is a false positive (Colquhoun 2017)[11]

Colquhoun is not alone. The American Statistical Association (ASA) has been stirred into action, and released a set of principles for the use and interpretation of p-values (Wasserstein and Amer Statistical Assoc 2016). Of these principles, the following are of particular relevance to CAM research.[12]

1. *P-values do not measure the probability that the studied hypothesis is true, or the probability that the data were produced by random chance alone.*
2. *Scientific conclusions should not be based only on whether a p-value passes a specific threshold.*
3. *By itself, a p-value does not provide a good measure of evidence regarding a model or hypothesis.*

These principles are so important in the context of CAM that some elaboration will be of value, by way of a hypothetical example. Dr. Vilseledda is a diligent but deluded physician who sincerely believes that remotely praying for unknown patients is likely to yield a clinical benefit.[13] He sets out to 'prove' this by way of a prospective clinical trial, where the subjects are terminally ill cancer patients assessed to have only 50% chance of surviving for one year. The primary outcome measure of the trial is 'percentage of patients alive after 12 months'. Dr. Vilseledda wants to do his research correctly, and so consults a statistician, Prof. Klokt, at the design stage, asking "what sample size do I need?" Prof. Klokt replies, "first I need to know what *effect size* you are expecting". Dr. Vilseledda is confident that prayer is an effective treatment, and so estimates that its use will give a 70% survival rate at one year. Prof. Klokt also wants to know what *power* is wanted; Dr. Vilseledda plumps for 80%, knowing that this is a typical and respectable figure in clinical trials. Prof. Klokt computes these values and advises that a sample size of at least 186 subjects is required.[14]

So, Dr. Vilseledda obtains the paper medical records of 200 terminally ill cancer sufferers who fit his clinical criteria (none of which are his own patients), and has an

---

[11]An alternative way of expressing this is to say that if you observe $p = 0.05$ then, in order to achieve a false positive risk of 5%, you would need a prior probability of 87%—clearly pre-posterously high (Colquhoun 2017).

[12]The numbering is ours, and the wording of #2 had been adapted slightly, to remove reference to 'business or policy decisions'.

[13]This example is chosen to illustrate an inherently absurd modality, recognisable as such to all reasonable people. Sadly however, proponents of such 'intercessory therapeutic prayer' exist; indeed, some of them have even conducted 'clinical trials' into this form of CAM (Roberts et al. 2009). We shall consider one such real-life case later in the next chapter.

[14]An online statistics tool was used to calculate this sample size (ClinCalc LLC 2017).

independent colleague randomly divide them into a 'prayer' group (n = 100) and 'control' group (n = 100). None of the patients are aware which group they have been assigned to, nor do these patients have any interaction with Dr. Vilseledda or anyone else involved in the study. A priest is asked to spend 10 min praying, 3 times per week for 6 weeks, over a box containing the medical records of those in the prayer group, urging God to heal these patients. Meanwhile, the medical records of the control group are simply placed in an identical box and not subject to prayer. At 12 months following the conclusion of the prayer treatment, Dr. Vilseledda has an independent colleague collect hospital data showing how many patients in each group are still alive.

Let us suppose that the results are as follows:

- Control group: 47% alive
- Prayer group: 61% alive.

Dr. Vilseledda is very excited about this result, because on first sight it appears to support his belief that intercessory prayer can be an effective therapy. However, he knows that the difference between the two groups' survival rates needs to be 'statistically significant', before any journal will publish his findings. So, he again consults Prof. Klokt, who performs a significance test on the data. This yields a p-value of 0.0466.[15]

Dr. Vilseledda is delighted. A p-value of 0.0466 is 'statistically significant', using the commonly accepted threshold of $p = 0.05$. What's more, Dr. Vilseledda knows that his was a well-designed study: it was randomized and double-blinded; it looked forward in time; it had a long duration for data collection; it had adequate power; and no participants dropped out (indeed they were never even aware of being on the trial![16]). So, he writes a paper based on this research, titled 'Intercessory prayer significantly increases the survival of cancer patients', which is duly submitted for publication in an academic journal.

When Prof. Klokt hears of this, she is perplexed, and challenges Dr. Vilseledda, saying "you have no basis on which to make such a claim". Dr. Vilseledda protests: "but $p = 0.0466$ shows that there is less than a 5% chance that my data were obtained by random chance alone. In other words, there is a greater than 95% probability that my belief in intercessory prayer is true!" But, as Prof. Klokt points out, this is in violation of the ASA principles referred to above. Prof. Klokt starts with this principle:

1. *P-values do not measure the probability that the studied hypothesis is true, or the probability that the data were produced by random chance alone.*

---

[15]This was calculated using a z-test; other suitable tests exist, but all of these will approximate to this value.

[16]There is an interesting ethical side question here: should subjects who cannot feasibly be affected in any way whatsoever by a remote experiment (such as intercessory prayer) be required to give consent?

Prof. Klokt cites the following quote (from above) to Dr. Vilseledda: "...*if we observe a P value just below 0.05, then there is a chance of at least 26% that your result is a false positive*". She further points out that intercessory prayer is a wholly implausible modality, meaning that the prior probability of positive effects should be considered as being extremely low. For instance, if the prior probability is 0.01%, this study would (per the logic described above) be expected to give a false positive risk of well over 99%. Moreover, most reasonable persons would agree that intercessory prayer simply cannot have any clinical effects: this equates with a prior probability of 0%, which in turn gives a false positive risk of 100%.

Prof. Klokt points out to Dr. Vilseledda that, by considering only $p = 0.0466$ and failing to factor in plausibility, he has also violated this ASA principle:

2. *Scientific conclusions should not be based only on whether a p-value passes a specific threshold.*

And finally, Prof. Klokt rebukes Dr. Vilseledda for violating this ASA principle:

3. *By itself, a p-value does not provide a good measure of evidence regarding a model or hypothesis.*

Here, Dr. Vilseledda has made the mistake of assuming that $p = 0.0466$ provides good evidence to support his belief that 'intercessory prayer can improve survival'. But this is only one possible explanation that is compatible with the results. A competing explanation is that the results were due to chance alone. In science, where two (or more) competing explanations exist, and both fit the data equally well, an intellectual tool known as 'Occam's razor' is used to decide which explanation to accept (at least until further data become available). Occam's razor requires that the explanation with the fewer assumptions ought to be preferred.

Applying Occam's razor to Dr. Vilseledda's study, it is obvious that the inter-cessory prayer explanation contains a number of major assumptions, including the existence of an omnipresent supernatural being who can detect people's prayers and intercede to perform miracles. By contrast, the competing explanation contains only one assumption, and a modest one at that: namely that a cluster of longer-lived patients happened by chance to have been assigned to the prayer group.

2.5 Statistics: Conclusions

The $p = 0.05$ threshold is often misunderstood and misinterpreted in science, leading to inaccurate analyses of data. In clinical research in general, and CAM in particular, this frequently leads to erroneous, inflated claims of clinical effectiveness.

Short of adopting a fully Bayesian approach to statistics, what should medical researchers do? One simple suggestion has recently been put forward by a team of statisticians: change the default P-value threshold for statistical significance for claims of new discoveries from $p = 0.05$ to $p = 0.005$ (Benjamin et al. 2017). This more demanding threshold would add a safety margin that would be effective for most trials. However, this solution applies only where the therapy under test has some plausibility, meaning a prior probability substantially greater than zero (generally no less than 3%, as an absolute minimum). And for therapies with vanishingly small (or zero) prior probabilities, even this threshold will be insufficiently stringent.

Another solution is suggested by David Colquhoun: the terms "significant" and "non-significant" should simply never be used (Colquhoun 2017). His recommended alternative approach is that p-values should be supplemented by specifying the prior probability that would be needed to produce a specified (e.g. 5%) false positive risk. For example, if we observe $p = 0.05$ then, in order to get a false positive risk of 0.05 one would need to postulate a prior probability (that the effect was really not zero) of 87%. In other words, one would need to be almost sure that there was a real effect *before* doing the experiment. (Again, this shows the weakness of the evidence provided by a p-value close to 0.05.)

It is important to realize that the foregoing criticisms of frequentist statistics, significance testing, and p-values are made in the context of inherently well-designed experimental studies. So, in our hypothetical example of an intercessory prayer trail, Dr. Vilseledda's design had no obvious structural deficits, and it was well executed. By contrast, as will become apparent in the next chapter when we examine real life cases, CAM trials are frequently very far removed from any reasonable definition of robust design. Common flaws include lack of sufficient numbers of subjects (leading to under-powered trials), no true control group, no placebo, lack of randomization, and no double-blinding. Moreover, data analysis in CAM is frequently disingenuous, with various forms of bad practice commonly employed, including the use of surrogate endpoints and data-dredging in search of anything 'statistically significant'. Added to these defects and poor practices are the less visible (but undoubtedly common) factors of sloppy trial management (for example a breakdown of double-blinding) and outright fraud. When the design or execution of a study, or the analysis of its data, is seriously flawed, it does not matter whether the statistical cut-off is $p = 0.05$ or $p = 0.005$ or for that matter $p = 0.0005$: the results should not be taken seriously; and for researchers (or anyone else) to do otherwise is ethically questionable.

3   Systematic Reviews

Even large-scale, high quality RCTs are rarely 'perfect'. RCTs seek to discover statistical correlations between a treatment and medically relevant changes in patients. As explained above, even 'statistically significant' positive results from a well-designed trial give no guarantee that the results did not occur by chance—in other words, that the correlation is spurious. This is particularly so in RCTs which yield only modest correlations, or marginal subject responses. Thus, a single clinical trial, even if well-designed, will not be adequate for establishing clinical effectiveness beyond doubt. Unfortunately, multiple trials are rarely in full agreement with each other: once again the 'multiple comparisons' problem is important here, with outliers (i.e. false-positives) present amongst published reports. In such cases, the temptation amongst CAM proponents to 'cherry-pick' the outliers can be irresistible.

Consider, for instance, the case of homeopathy as a treatment of hay fever. Let's assume there are 12 RCTs on the subject; 3 suggest that homeopathy is efficacious, and 9 imply it does not work better than a placebo. In such a situation, proponents of homeopathy could easily select the 3 positive trials to claim that their treatment works, while sceptics could use the 9 other studies to show exactly the opposite. Both would be arguing based on RCT evidence and would seemingly be correct. Yet both cannot be correct!

To get closer to the truth, some researchers specialise in methodically combining the results of all published studies, including but often not limited to RCTs, and generating an overall conclusion as to the effectiveness of the treatment in question. The type of publication resulting from such research is known as a 'systematic review'. High-quality systematic reviews conform to well-established rigorous methods for selecting all relevant published studies, evaluating their trustworthiness and statistically pooling their results.[17] The best reviews conform to well-established PRISMA guidelines, which set out criteria and standards for the correct conduct of systematic reviews (Moher et al. 2009). Because the systematic review approach combines the data from several independent RCTs, it can be a much better way to answer clinical questions than relying on any one individual study.

Another limitation with RCTs is that it is never possible to establish the safety of a therapy based on a few clinical trials. To assess therapeutic safety, one would need sample sizes that go two or three dimensions beyond those of RCTs. Thus, safety assessments are best done by evaluating the evidence from all available credible sources, including epidemiological investigations and observational studies. By combining and triangulating all such relevant evidence, the systematic review approach offers a reasonable means to ascertain therapeutic safety. However, RCTs of CAM frequently fail to mention adverse effects, which is not only unethical per se, in terms of withholding information that could prevent harm, but also contributes towards a false positive picture when systematic reviews are generated based on such data.

Despite their intrinsic strengths, systematic reviews are often not 'perfect'. This is so for several reasons. First, systematic reviews may themselves be poorly conceived or improperly conducted; for example, there may be a failure to apply strict exclusion criteria, leading to data from very weak studies being considered, thus undermining the systematic review's conclusions. Second, it is difficult, and

---

[17]The statistical analysis of such combined data is referred to as 'meta-analysis', and some systematic review publications are titled as meta-analyses.

often impossible, to discern amongst published studies those that are of acceptable quality from those that, while their design is acceptable, have been sloppily conducted, or are fraudulent. Third, in addition to (or instead of) examining RCTs, some systematic reviews looking at clinical efficacy include intrinsically weaker forms of research, such as observational studies. This may be because in each area there is a dearth of published RCT research—a problem that afflicts many areas of CAM. As discussed above, data from observational studies cannot demonstrate cause and effect. Fourth, some reviewers may deliberately omit primary data that does not fit their aspirations.

A further limitation of systematic reviews stems from the fact that 'positive' trials—i.e. those that show a positive correlation between a treatment and a beneficial outcome—are more likely to get published than their 'negative' counterparts. Several causes underlie this phenomenon, which is known as 'publication bias'. Most simply, researchers may fail to write up and publish negative trial results; for instance because they wish to forget about the 'failed' trial and move on to some other work. A related problem is that editors of medical journals, and the peer reviewers responsible for assessing submitted articles for publication, tend to favour 'positive' results and reject 'negative' ones—perhaps because the former seem more 'interesting' than the latter. This is particularly marked amongst CAM journals, where it is rare to fine negative results published at all (Ernst and Pittler 1997). As well as directly skewing publication in the 'positive' direction, knowledge of the existence of this tendency serves to further discourage researchers from attempting to publish negative results. Non-publication of negative results can also be due to a researcher deliberately choosing to 'bury' their negative results, either because they do not believe that a set of results is 'real', or because they actively (and fraudulently) wish to hide the fact that their favoured treatment may be useless. Whatever the motivations of the individuals concerned, the net result of publication bias is that the overall landscape of published clinical trials gets skewed towards positive reports.

Because authors of systematic reviews are often forced to fish for papers amongst this sea that is infested with an excess of false-positive reports, the resulting reviews are themselves prone to being skewed in a positive direction. The risk of this happening is dependent upon the area in question: fields that attract sufficient high quality research have a relatively low proportion of false-positive publications and therefore a relatively low risk of generating erroneously positive systematic reviews. By contrast, any field in which sloppy and fraudulent research is more common will tend to generate systematic reviews that present a grossly inaccurate or false picture of the apparent effectiveness of the treatments concerned.

Regrettably, as we will show in the next chapter, research within many areas of CAM is of lamentably low quality and reliability. Moreover, the resultant skewing of systematic reviews naturally tends to be exacerbated where the systematic review authors are themselves part of a research field that has an endemic culture of slackness and deception, because such authors are likely to apply the same problematic mind-sets and approaches to the systematic review process. Scientists who are independent of the field concerned generally make a better job of such systematic reviews; however, even here the scope for truth is limited if the nature of the published body of primary work is grossly flawed.

So far in this chapter, we have set out the features of good and bad clinical research. In this context, we need to ask: in general, how does CAM research measure up? We have already alluded to weak and unethical research, and we claimed that CAM research has more than its fair share of such malpractice. We shall now consider this question in more detail.

4 Research Activity: Mainstream Versus CAM

For some time now, the research activity in CAM has been seemingly buoyant. In recent years, Medline (the largest database of medical articles) listed several thousand papers in the category of 'complementary alternative medicine'. This may seem impressive. Yet, compared to other areas of medical research, this figure is modest. Here are numbers of Medline-listed articles published in 2015 for a few major mainstream fields:

- Internal medicine: 36,998
- Paediatrics: 30,646
- Pharmacology: 194,322
- Psychology: 65,679
- Surgery: 176,277.

But what about specific CAM modalities? Here are the numbers of Medline-listed articles published in 2015 for some of the major CAM treatments:

- Acupuncture: 1784
- Chiropractic: 314
- Herbal medicine: 1572
- Homeopathy: 181
- Naturopathy: 45.

These figures could show that, for some areas of CAM, there are more research funds and/or expertise than for others. Alternatively, they might indicate that certain sections of CAM are more open to scientific scrutiny than others. A closer look at the research activity in defined CAM therapies might create more insights.

For example, in the case of homeopathy, further analysis led to the following conclusions (Ernst 2016b):

- The research activity into homeopathy is currently very subdued.
- The arguably most important research question of effectiveness attracts very little research.
- There is a relative abundance of papers that are data-free opinion pieces.

And in the case of acupuncture, the conclusions were (Ernst 2016a):

- Too little research is focused on the two big questions: effectiveness and safety.
- In relation to the meagre output in RCTs, there is a disproportionate number of review articles.
- The majority of the articles were published in low or very low impact journals.
- Papers in 'peer-reviewed' CAM journals will usually have been reviewed by CAM sympathizers.

These findings, even though based on very cursory analyses, seem to imply that CAM research is markedly different from that in other areas of medical research—and not in desirable ways. CAM research is currently far less active than other areas of medical investigation. Where such research is conducted, the arguably most important medical research questions, namely efficacy and safety, do not seem to be a priority to CAM researchers. In fact, most of the articles being published are data-free and thrive on mere opinion. And of the papers that do contain empirical data, many are based on surveys of patients.

Patient surveys can be of genuine value, of course. For example, it might be useful to survey patients who undergo surgical pregnancy termination, to determine which pre-surgical interactions and environments best relax the patient. However, while surveys contribute a small minority of the research publications in mainstream healthcare, survey-based papers dominate the field of CAM research: there is no field of medical research that produces more surveys than CAM. Around 500 CAM surveys are being published every year; this amounts to about two every working day which is substantially more than the number of clinical trials in this field.

This emphasis on survey-based research by CAM researchers can be accounted for by several factors:

- Surveys can be done without substantial funding.
- They generate publishable data much quicker than clinical trials.
- They do not require much research expertise.
- They almost invariably produce results that can be used for the promotion of CAM.

The typical survey is based on questionnaires which are distributed to patients asking them whether they employ any form of CAM, for what condition and with what effect. Such surveys typically show that:

- A high percentage of patients use CAM.
- They employ it mostly for benign, self-limiting conditions.
- Users of CAM are mostly satisfied with their choices.
- Most CAM users are well-educated, affluent, middle-aged and female.

Such insights might be interesting to note once or twice, but not hundreds of times. Moreover, it needs to be noted that most of these surveys are methodologically so weak that they contribute virtually nothing to our knowledge. These are good reasons to be critical of this 'survey-mania' (Ernst 2006). For a typical prevalence survey, a team of enthusiastic researchers might put together a few questions and design a questionnaire to find out what percentage of a group of individuals have tried CAM in the past. Subsequently, the investigators might get one or two hundred responses. They then calculate simple descriptive statistics and demonstrate that xy% (let's assume it is 45%) use CAM. This finding eventually gets published in one of the many CAM journals, and everyone is happy— well, almost everybody. In most cases, this is not research, it is pseudo-research which ignores most of the principles of survey-design. The typical CAM prevalence survey has none of the features that would render it a scientific investigation:

(1) It lacks an accepted definition of what is being surveyed. There is no generally accepted definition of CAM, and even if the researchers address specific therapies, they run into huge problems. Take prayer, for instance—some see this as a form of CAM, while others would, of course, argue that it is a religious pursuit. Or take herbal medicine—many consumers confuse it with homeopathy, some might think that drinking tea is herbal medicine, while others would probably disagree.

(2) The questionnaires used for such surveys are almost never validated. Essentially, this means that we cannot be sure they evaluate what the authors think they evaluate. We all know that the way we formulate a question can determine the answer. There are many potential sources of bias here, and they are rarely taken into consideration.

(3) Enthusiastic researchers of CAM usually recruit a small convenience sample of participants for their surveys. This means they ask a few people who happen to be around to fill their questionnaire. Consequently, there is no way the survey is representative of the population in question.

(4) The typical survey has a low response rate; sometimes the response rate is not even provided or remains unknown even to the investigators. This means we do not know how most patients/consumers who received but did not fill the questionnaire would have answered. Often there is good reason to suspect that those who have a certain attitude did respond, while those with a different opinion did not. This self-selection process is likely to produce misleading findings (Ernst 2013a).

In our view, the plethora of CAM surveys is counter-productive. This is so for several reasons: they tend to grossly over-estimate the popularity of CAM; they distract money, labour and attention from the truly important research questions in this field; and they give a false impression of a buoyant research activity. Crucially, we should know that the unreliable findings of such investigations are frequently used to mislead the public. The fallacious logic often is as follows:

1. May people use CAM and pay for it out of their own pockets.
2. Most are very satisfied with the effects of the CAM therapies of their choice.
3. This means CAM is effective.
4. It is unjust that only people who can afford the expense can benefit from CAM.
5. CAM should therefore be made available to everyone who wants it free of charge.

5   Problems with CAM Research

Although most of the published CAM articles are merely opinion, speculation or survey-based, a number of empirical studies aimed at examining the efficacy of CAM have emerged. Earlier in this chapter we pointed out that, despite various possible safeguards against bias, clinical trials of CAM modalities are often less than rigorous and their findings less than reliable. There are several factors that can contribute to this phenomenon. We have already mentioned that high quality clinical research tends to be expensive, and research funds are particularly scarce in CAM. To this we can add the following two factors:

- The minimization of bias in research requires expert scientists, and very few such individuals feel attracted to CAM. In fact, CAM is often seen by young talent as a means to ruin a promising career.
- Researchers of CAM are often enthusiasts of the treatments they investigate. Consequently, bias is often allowed to creep into their studies.

The last point seems particularly important. It is possible to identify a disappointingly long list of CAM researchers who rarely or never publish a negative finding related to 'their' therapy. This means that, in the hands of such 'experts', the treatment in question is efficacious in virtually every situation. Similarly, it has been shown repeatedly that studies of TCM from China hardly ever report negative results (Tang et al. 1999; Vickers et al. 1998), and that data fabrication in China is an 'open secret', and particularly affects CAM (Ernst 2016c). Such phenomena cast considerable doubt on the trustworthiness of the research published.

Many independent observers have remarked that much of the research in CAM is far from rigorous. In fact, we would go one step further and claim that many studies in this area actively mislead the public about the value of CAM. As this may seem like a bold statement, in the following chapter, we intend to provide evidence for it focusing on examples of clinical research addressing the two major questions of efficacy and safety of CAM therapies. Most of the examples we give contain more than one flaw; indeed, the careful reader will doubtlessly be able to spot many of the violations of the principles of ethical medical research we discussed above. However, it would be beyond the scope of the present chapter to detail each and every flaw in examples of CAM research—there are simply too many! Instead we shall focus primarily on the following issues:

- Insufficient numbers of participants
- Lack of a control group
- Inappropriate control group
- Control group present but otherwise flawed
- Use of surrogate endpoints
- Misuse of statistics
- 'Too good to be true' results
- Fraudulent research.

In the next chapter, we shall consider many real-world instances of unsound CAM research. Before doing so, we shall briefly address one question that may be troubling readers: why does inherently flawed CAM research end up being published? To be published in an academic journal (in any field), research papers must pass an initial evaluation by an editor and then be scrutinised by (typically two) peer reviewers. At first glance, this time-honoured method ('peer-review') of filtering research papers would seem to be rigorous, and ought to serve as an effective screen for quality. Indeed, if editors and peer reviewers are diligent, well-informed and unbiased, this method can ensure that flawed papers are not published. However, this system can break down in several ways. Editors or—especially—reviewers who are rushed or insufficiently expert in the field concerned can fail to identify problems. The issue of non-diligent, non-expert reviewers is a problem in academic publishing in general, and is exacerbated by the fact that the present model of peer review is based on altruism. In other words, peer reviewers are not recompensed for their work. This can lead to reviewers devoting insufficient time to the task. Another consequence of the *pro bono* expectation is that busy researchers are frequently unwilling to accept journals' requests to review; thus, journals often have no option but to give the task to reviewers who are willing but of questionable expertise.

While the above problems are by no means restricted to CAM journals, these journals are inherently more susceptible to these problems than are their mainstream medical counterparts. This is because few competent scientists are willing to spend valuable time reviewing CAM research. Occasionally, it even seems doubtful whether CAM journal editors really want qualified unbiased scientists to scrutinise submitted papers: if such willing scientists can be found, the most probable outcome of their scrutiny will be rejection! But editors need to fill the pages of their journals. In reality, CAM journals allocate peer review tasks to a narrow range of CAM enthusiasts who often have been chosen by the authors of the article in question. The raison d'être of CAM journals and CAM researchers is inextricably tied to a belief in CAM, resulting in a self-referential situation which is permissive to the acceptance of weak or flawed reports of clinical effectiveness.

6 Conclusions

Research is crucial in clinical medicine: it is how knowledge of treatment effectiveness and safety is gained and how progress is made. It is essential that clinical research is of the highest quality in terms of its design, execution and communication. In recent decades, mainstream medical science has travelled away from a reliance on weak forms of evidence, such as anecdotal accounts, towards rigorous experimentally based clinical studies. In terms of study design, the RCT is the generally-accepted gold-standard for testing the efficacy of treatments. All quality RCTs possess certain key design features in their protocol and execution: sufficient numbers of patients are recruited; these subjects are randomly assigned to control and experimental groups; all participants in the study—subjects and researchers alike—are 'blinded' as to which patients receive the test treatment and which receive the placebo; and the trial duration is of suitable length to properly assess treatment responses.

Yet, despite the great progress that has been made in improving research design, many CAM enthusiasts prefer to remain in the epistemic dark ages, advocating that the best source of evidence of clinical effectiveness resides in traditional knowledge and subjective observations by the practitioner.

Nevertheless, some CAM researchers appear to see the need to go beyond merely tradition and anecdotes, adhering to the notion that established scientific approaches to research provide the best (or at least a valid) means to obtaining reliable clinical knowledge. Sadly, however, when published papers from CAM clinical trials are scrutinised, all too frequently a common problematic pattern is detected. On first sight, the study appears to conform to the norms of good medical research, but on closer inspection it becomes apparent that this appearance of scientific rigour is merely a veneer hiding the truth—that the study was deeply flawed in some key respect or other. Moreover, the statistical treatment of data in CAM studies often turns out to be fundamentally flawed.

It is crucial that data from clinical studies are carefully analysed using valid statistical approaches. Dismayingly, it has become evident that—even in mainstream medical research—such analysis has frequently been based upon flawed assumptions, contributing to the 'reproducibility crisis' that science in general is presently experiencing. Nevertheless, *bona fide* researchers are gradually learning to improve their approaches to interpreting data, and in this way science is displaying one of its hallmark characteristics, namely that of self-correction. However, the world of CAM research is lagging in these developments. While many academic papers, Internet articles and blog postings discussing these statistical issues have recently been published by medical scientists, the CAM community has been almost silent on the subject.[18]

Defective research—whether at the design, execution, analysis, or reporting stage—corrupts the repository of reliable medical knowledge. Ultimately, this leads to suboptimal and erroneous treatment decisions. As will become clear in the next chapter, when we examine a wide range of typical CAM research publications, this kind of ethically reprehensible approach is commonplace in CAM.

---

[18]For example, a search of the academic publications database Web of Science (which includes all the major CAM journals) covering 2007–2017 using various relevant search terms (p-value, reproducibility crisis, p-hacking) led to merely one CAM paper dealing with the issue of problematic statistical interpretation of clinical data (Benbassat 2016).

References

Beecher H (1955) The powerful placebo. J Am Med Assoc (JAMA) 159(17):1602–1606

Benbassat J (2016) Inferences from unexpected findings of scientific research: common misconceptions. Eur J Integr Med 8(3):188–190. doi:10.1016/j.eujim.2015.12.010

Benjamin DJ, Berger J, Johannesson M et al (2017) Redefine statistical significance. PysArXiv preprints. https://osf.io/preprints/psyarxiv/mky9j/. Accessed 14 Aug 2017

Benjamini Y, Hechtlinger Y (2014) Discussion: an estimate of the science-wise false discovery rate and applications to top medical journals by Jager and Leek. Biostatistics 15(1):13–16. doi:10.1093/biostatistics/kxt032

ClinCalc LLC (2017) Sample size calculator. ClinCalc.com. http://clincalc.com/stats/samplesize.aspx. Accessed 11 Feb 2017

Colquhoun D (2014) An investigation of the false discovery rate and the misinterpretation of p-values. Roy Soc Open Sci 1(3):140216. doi:10.1098/rsos.140216

Colquhoun D (2017) The reproducibility of research and the misinterpretation of P values. Roy Soc Open Sci. http://www.biorxiv.org/content/biorxiv/early/2017/08/07/144337.full.pdf. Accessed 12 Aug 2017

Ernst E (2006) Prevalence surveys: to be taken with a pinch of salt. Complement Ther Clin Pract 12(4):272–275. doi:S1744-3881(06)00042-9 [pii]

Ernst E (2013a) More dismal chiropractic research. Edzard Ernst | MD, PhD, FMedSci, FSB, FRCP, FRCPEd. http://edzardernst.com/2013/02/more-dismal-research-of-chiropractic/. Accessed 15 Jan 2017

Ernst E (2013b) What can be more irresponsible than implying that homeopathy cures cancer? Edzard Ernst | MD, PhD, FMedSci, FSB, FRCP, FRCPEd. http://edzardernst.com/2013/09/what-can-be-more-irresponsible-than-implying-that-homeopathy-cures-cancer/. Accessed 15 Jan 2017

Ernst E (2016a) Acupuncture: the current state of the scientific literature. Edzard Ernst|MD, PhD, FMedSci, FSB, FRCP, FRCPEd. http://edzardernst.com/2016/03/acupuncture-the-current-state-of-the-scientific-literature/. Accessed 15 Jan 2017

Ernst E (2016b) The current state of research into homeopathy. Edzard Ernst | MD, PhD, FMedSci, FSB, FRCP, FRCPEd. http://edzardernst.com/2016/08/the-current-state-of-research-into-homeopathy/. Accessed 15 Jan 2017

Ernst E (2016c) Data fabrication in China is an 'open secret'. Edzard Ernst | MD, PhD, FMedSci, FSB, FRCP, FRCPEd. http://edzardernst.com/2016/10/data-fabrication-in-china-is-an-open-secret/. Accessed 15 Jan 2017

Ernst E, Pittler M (1997) Alternative therapy bias. Nature 385(6616):480. doi:10.1038/385480c0

Hrobjartsson A, Gotzsche PC (2010) Placebo interventions for all clinical conditions. Cochrane Database of Syst Rev (1):CD003974. doi:10.1002/14651858.CD003974.pub3

Ioannidis J (2005) Why most published research findings are false. Plos Medicine 2(8):696–701. doi:10.1371/journal.pmed.0020124

Ioannidis J (2014) Discussion: why "an estimate of the science-wise false discovery rate and application to the top medical literature" is false. Biostatistics 15(1):28–36. doi:10.1093/biostatistics/kxt036

Jager LR, Leek JT (2014) An estimate of the science-wise false discovery rate and application to the top medical literature. Biostatistics 15(1):1–12. doi:10.1093/biostatistics/kxt007

McGorry R, Webster B, Snook S et al (2000) The relation between pain intensity, disability, and the episodic nature of chronic and recurrent low back pain. Spine 25(7):834–840. doi:10.1097/00007632-200004010-00012

Medawar PB (1967) The art of the soluble. Methuen, London

Moher D, Liberati A, Tetzlaff J et al (2009) Preferred reporting items for systematic reviews and meta-analyses: the PRISMA statement. Plos Med 6(7):e1000097. doi:10.1371/journal.pmed.1000097

Nuzzo R (2014) Statistical errors. Nature 506(7487):150–152

Pandolfi M, Carreras G (2014) The faulty statistics of complementary alternative medicine (CAM). Eur J Intern Med 25(7):607–609. doi:10.1016/j.ejim.2014.05.014

Roberts L, Ahmed I, Hall S et al (2009) Intercessory prayer for the alleviation of ill health. Cochrane Database Syst Rev (2):CD000368. doi:10.1002/14651858.CD000368.pub3

Schwalbe M (2016) Statistical challenges in assessing and fostering the reproducibility of scientific results. National Academy of Sciences, USA

Tang JL, Zhan SY, Ernst E (1999) Review of randomised controlled trials of traditional Chinese medicine. BMJ 319(7203):160–161

Vickers A, Goyal N, Harland R et al (1998) Do certain countries produce only positive results? A systematic review of controlled trials. Control Clin Trials 19(2):159–166. doi:10.1016/S0197-2456(97)00150-5 [pii]

Wasserstein RL, Amer Statistical Assoc (2016) ASA statement on statistical significance and P-values. Am Stat 70(2):131–133

Chapter 3
The Reality of CAM Research

In this chapter, we shall explore the various ways in which CAM research frequently falls short of the ideals of good research practice, using several examples of seriously flawed CAM research studies. We will start with the least severe forms of research misconduct, in terms of the moral culpability of the perpetrators, and progress to the most blatantly corrupt and unethical practices. The chapter will conclude with some broad questions about the ethics of doing research in CAM.

1 Problems with Controls

As we explained in the previous chapter, the presence of a suitable control group is crucial in any clinical trial. But published CAM studies frequently breach this principle and often have no control group at all. The following examples, which cover a range of different CAM modalities, illustrate this fundamental omission. It is important to stress: even though we can only employ a few examples here, there are many such CAM studies. In fact, it seems hard to find one which is not burdened with major limitations of the kind discussed below.

1.1 No Control Group: Reiki and Stress Management

We shall start with a study intended to determine the effects of a Reiki programme designed to reduce stress and induce relaxation. Reiki is a CAM therapy in which the practitioner's hands hover above, or occasionally lightly touch, the patient's body. According to Reiki's proponents, this induces a 'healing energy' to flow within the patient's body, resulting in therapeutic benefits. From a plausibility standpoint, this is of course highly problematic: despite many decades of physiological research, no such 'energy' has ever been discovered by scientists. Even if

© Springer International Publishing AG 2018
E. Ernst and K. Smith, *More Harm than Good?*
https://doi.org/10.1007/978-3-319-69941-7_3

this 'energy' did exist, there is no rational or scientific basis to believe that its 'flow' could be improved by a Reiki practitioner's hands. Undeterred by this plausibility vacuum, researchers have attempted to evaluate the clinical effectiveness of Reiki, as in our example below.

This study investigated the use of a form of Reiki for stress management (Bukowski 2015). Students were recruited for a 20-week structured self-Reiki program, with twice weekly sessions being conducted in the privacy of their residence. Each volunteer completed symptom score questionnaires every 4 weeks. Twenty participants completed the study; apart from two students, the participants believed that Reiki would be effective in reducing stress levels. The study found a significant reduction in average stress levels as measured by symptom scores. And, with one exception, stress levels at 20 weeks did not return to pre-study stress levels. The authors concluded that "*this study supports the hypothesis that the calming effect of Reiki may be achieved through the use of self-Reiki. Self-Reiki is one potential stress-management method that is applicable to college students*".

This investigation looked for evidence of a relaxation effect and an attendant reduction in stress, as opposed to seeking improvements in a pathological condition. This is a relatively modest claim for any therapy, and one which is commonplace in many CAM modalities. Yet, based on their findings, the authors were not entitled to make any such claim, because their study is fatally flawed. Prominent amongst several defects is the lack of a control group. This omission means that there are no data against which to compare the recorded improvements in stress, which may have come about for reasons that have nothing to do with 'life energy' or Reiki. The reported reduction in stress could have been a result of several possible factors, including: the get-better-anyway effect; a desire to please the researchers; experimenter-expectancy effects, associated with prior faith in 'mind-body' therapies; a placebo response elicited via the attention received by enrolling on the study; and cognitive errors on the part of the researchers such as expectation bias and wishful thinking.

Thus, from this study, it is impossible to know whether the Reiki treatment investigated had any specific therapeutic benefits at all. For the research to have been valid, a proper control group would have been required, for instance, comprising volunteers assigned to a 'sham Reiki' program, entailing non-Reiki intervention. Only if the true Reiki group showed a significantly higher improvement in stress than the sham group, could it be concluded that the Reiki treatment may have had any specific benefits.

An additional, obvious flaw of this study is that it is far too small to generate findings that would allow any valid conclusion to be drawn as to therapeutic effectiveness. This would have remained the case even if a control group had been employed, save in the (highly implausible) scenario in which the Reiki treatment vastly out-performed the sham treatment.

In other words, this paper misleads uncritical readers into believing in, and possibly adopting, a therapy that is not supported by good evidence. This could lead individual patient-consumers, or decision-makers in healthcare, to waste time and money on an ineffective therapy, when effective forms of stress management are available. This is of obvious ethical concern.

There is a further, deeper issue here; of the myriad of conventional ways available that may reduce stress—going for a walk, socialising, massage, playing sport, listening to music, watching TV, running, or for more severe cases pharmaceutical medication—CAM therapies are different, in that they generally make claims that are pseudoscientific or magical. The 'life energy' claim central to Reiki (and indeed present in several other CAM modalities) is a good example of such a fantastical notion. Of all possible stress-reducing approaches, only CAM treatments depend upon the promulgation of such false beliefs. Voltaire made the point that 'those who make us believe in absurdities can make us commit atrocities'. In keeping with this dictum, we consider that any approach that relies on duping the recipient into believing fantasies is ethically unacceptable.

1.2 No Control Group: Homeopathy and Diabetes

Many CAM trials involve the treatment of serious pathological conditions, as opposed to investigating therapies intended simply to relax the patient. A study by the 'Indian Central Council for Research in Homeopathy' provides one such example (Nayak et al. 2013).

This clinical trial was aimed at evaluating homeopathic treatment in the management of diabetic polyneuropathy (DPN). This is a serious condition entailing damage to the patient's nervous system. In the study, the homeopathic medicine was selected and prescribed on an individualized basis. Patients were followed up for 12 months. On first sight, this trial may appear quite impressive, because a large number of participants were recruited: 336 diabetics initially enrolled in the study, of which 247 patients (those who attended regularly) were included in the final analysis. The main outcome measure analysed was a change in DNP symptoms. A statistically significant improvement in this parameter was found at 12 months. From this result, the authors concluded that "homeopathic medicines may be effective in managing the symptoms of DPN patients".

Despite its relatively large numbers of participants, this study is fatally flawed. As with our previous example, there is one overarching reason for this: there is no control group. The improvement in symptoms could be due to some patients administering conventional treatments for DNP, to a placebo-effect, or to the patients feeling better over time, or to any of the possible effects noted in our previous example. Without a control group to provide an adequate comparison, there is no way of knowing whether homeopathy had any specific effect at all. Accordingly, and contra to the authors' conclusion, the results from this trial are meaningless.

There is no good evidence here, or anywhere else, that homeopathy works for DPN. The conclusions of this paper might prompt patients to think otherwise. This is unethical as it offers false hope, potentially waste patients' (or public) resources, and put patients' health at risk by promulgating belief in an ineffective therapy, quite possibly at the expense of conventional therapeutic options. Accordingly, we consider this article to be an example of ethically unacceptable interpretation of data by CAM researchers.

1.3   No Control Group: Herbal Medicine and Sleep Quality

In our next example, the sedative effects of *Bryophyllum pinnatum*, a herbal medication predominantly used in a form of CAM known as 'anthroposophic medicine', were investigated (Lambrigger-Steiner et al. 2014). Seventy-eight pregnant women suffering from sleep problems were treated with *B. pinnatum*. Data were obtained for 49 of these participants. Sleep quality, daily sleepiness and fatigue were assessed with the aid of questionnaires, at the beginning of the treatment and after two weeks. The results showed no evidence for a prolongation of sleep duration, reduction in the time to fall asleep, nor improvement in fatigue severity. However, the number of wake-ups, and the subjective quality of sleep, were significantly improved. From these results, the authors concluded: "*B. pinnatum is a suitable treatment of sleep problems in pregnancy*".

Whether *B. pinnatum* really is a 'suitable treatment of sleep problems in pregnancy' simply cannot be determined from this trial, because no control group was included. Moreover, the plant contains cardiac glycosides which have the potential to cause serious adverse effects, and we know nothing about the remedy's potential to cause harm to the foetus.

Therefore, this study is even more problematic than our previous examples: its design is similarly flawed to the extent that its data are worthless; it fails to properly address potential health risks associated with the therapy; and it promulgates the unsubstantiated notion that *B. pinnatum* can improve sleep in pregnancy. In view of these facts, we consider this study to be ethically unacceptable.

1.4   No Control Group: CAM Medication for Cancer Hormone Therapy Side-Effects

In the previous examples, the findings were published in CAM journals of which most critical thinkers would automatically be sceptical. However, *bona fide* science-based journals are not immune from the risk of publishing flawed CAM research. An example is provided by the publication in the medical science journal In Vivo of a 'clinical investigation' performed '*to confirm the benefit of complementary medicine*' in patients with breast cancer undergoing hormone therapy

(HT) (Beuth et al. 2016).[1] A large number of patients (n = 1165) suffering from joint pain and mucosal dryness induced by the HT were enrolled and given a combination therapy comprising several CAM medicines.[2] Outcomes were documented before and 4 weeks after this treatment. Overall, most patients suffering from HT side effects reported a reduction in symptom severity. The reduction of side-effects of HT was statistically significant after 4 weeks. The authors concluded that their investigation showed that their combination of complementary medications "*significantly reduced defined side-effects (e.g. arthralgia, mucosal dryness) of adjuvant HT in patients with breast cancer.*"

This conclusion is not warranted by the data and is therefore dangerously misleading. As with the previous examples, this study had no control group. A whole host of factors other than the complementary medicines might have been responsible for the observed outcome, and the lack of a control group means that it is impossible to know whether the CAM therapy contributed to the observed improvements. As far as we can tell, the patients might even have fared better had they *not* taken the supplements!

But the authors apparently have no such concerns about the validity of their findings. This is evident when they go so far as to claim that, on the basis that tolerability to HT determines its optimal administration, the CAM treatment "*may enhance the chance of curing this disease*". We consider this to be an ethically reprehensible claim, particularly considering how understandably desperate many cancer patients are. It offers false hope to such patients, not only in terms of promoting an unproven remedy for reducing HT side effects, but—worse—by positing an unsubstantiated positive influence on the chances of the patient being cured of breast cancer.

Why did a mainstream journal publish such a flawed report? We cannot know for certain, but it is possible that some of the weaknesses inherent in the review process referred to in the previous chapter—such as the selection of non-diligent or non-expert reviewers—have played a part. In this regard, it may be of relevance that the In Vivo journal is of relatively low ranking (ranked 109 of 124 in the JCR Medicine, Research & Experimental category).[3] It is obvious that lower ranking journals have the greatest difficulty attracting both peer reviewers and high-quality papers. This does not mean that the In Vivo journal in general fails to properly review submitted papers, nor does it imply that most of the papers it publishes are flawed. Nor is it the case that low ranked journals in general necessarily carry lower quality papers; systems for ranking journals are imperfect, based as they are upon

---

[1]The stated aim of the In Vivo journal is thoroughly scientific; it is '*designed to bring together original high-quality works and reviews on experimental and clinical biomedical research within the frame of comparative physiology and pathology*'.

[2]The agents in the combination therapy were: sodium selenite (an agent popular amongst CAM advocates, despite very little evidence of in vivo benefits); bromelaine and papain (proteolytic plant enzymes used in herbal medicine and other CAM modalities); and Lens culinaris lectin (an extract from legumes commonly used in naturopathy).

[3]According to Journal Citation Reports® 2015 data.

numbers of citations of published papers. Anomalies certainly occur, such as specialist journals that publish extremely rigorous research paradoxically receiving a low ranking simply due to the highly specialist nature of their content attracting low numbers of citations. Nevertheless, as a generality, it is true that the risks of publishing seriously defective work are greater for lower ranked journals. Top ranked journals do also publish research that turns out to be badly flawed; however, this occurs at a lower frequency.

In all the above examples, no control group was included. As we discussed last chapter, all such studies are fatally flawed, because there is nothing to compare the results with. Sadly, studies that have no control arm are commonplace in CAM research.

Why do so many CAM studies exclude an appropriate control group? One possible answer is that it is simply easier to conduct a trial if it has no control group: only around half as many subjects need to be recruited; the necessity to create an appropriate control treatment (e.g. sham manipulations or similar-looking medicines) is obviated; and the inconvenience associated with randomisation and blinding is avoided. Considering that clinical research is intrinsically expensive and logistically difficult to carry out, and that CAM journals (and occasionally mainstream journals) are willing to publish the findings of uncontrolled studies, it is obvious that an incentive exists for CAM researchers to omit control groups.

Researchers who yield to the temptation of omitting a control group are either knowingly conducting futile and potentially misleading work, or they may simply lack the training, knowledge and competence required to conduct meaningful clinical research. In either case, a serious ethical problem exists. To knowingly engage in research that can only generate worthless results is a form of research malpractice. To carry out clinical research without having the requisite skills and knowledge is also ethically fraught: ignorance can be no defence, as researchers are morally obligated to ensure that they understand what they are doing.

Aside from CAM researchers being tempted to make life easier for themselves, a second possible reason for the omission of control groups is that the inclusion of an appropriate control treatment is very likely to generate results that show the CAM therapy to be ineffective. Consider for example homeopathy, and recall that homeopathic medicines are so diluted that typically they contain not a single molecule of the original 'active' substance. Apart for the scientifically deluded who actually believe that water retains a 'molecular memory' of the long-gone substance, it should be clear to any modestly intelligent CAM researcher that any responses in patients to a homeopathic medicine can only be due to non-specific effects, including the placebo response and the 'get better anyway' effect. Producing findings which demonstrate that a CAM therapy is no better than placebo is not an attractive proposition to CAM researchers who are ideologically wedded to a belief in the therapy under investigation. It is also the case that CAM journals do not like to publish such 'negative' findings. These factors may serve as a disincentive to some CAM researchers from making the effort to include a control group, as this would most likely undermine the hoped-for positive findings.

But not all CAM studies lack a control group; as we will see shortly, many published studies do include controls that provide comparison data. However, caution is required in interpreting CAM studies that contain such a group, as very often this is not a true control group at all—thus rendering the comparison invalid. The following four studies fall into this trap, in diverse ways.

1.5 Lack of an Adequate Control Group: Chiropractic and Acupressure for Headaches

Our first example in this category is a study authored by prominent US chiropractors and published in the Journal of Manipulative and Physiological Therapeutics, the leading chiropractic publication (Vernon et al. 2015). Its aim was to determine if the addition of a 'self-acupressure pillow' (SAP) to typical chiropractic treatment generates significantly greater improvement in headache sufferers than chiropractic care alone. Thirty-four patients were divided into two groups. Group A (n = 15) received typical chiropractic care only (manual therapy and exercises), and group B (n = 19) received the same chiropractic care plus daily home use of the SAP. The intervention period was 4 weeks. The principal outcome measure was headache frequency. The main findings were that group A obtained a 46% reduction in headache frequency, and the proportion of subjects in group A achieving a reduction in headaches to a threshold level[4] was 71%, while for group B it was 28%. Additionally, the mean satisfaction rating (on a scale of 0–3) of users of the SAP was 2.7. The authors concluded that "*this study suggests that chiropractic care may reduce frequency of headaches in patients with chronic tension-type and cervicogenic headache. The use of a self-acupressure pillow may help those with headache and headache pain relief as well as producing moderately high satisfaction with use.*"

Virtually none of these conclusions are supported by the actual data. Crucially, the effectiveness of either a self-administered acupressure cushion or chiropractic care alone remains unproven, because a suitable control group was not included. The study compared [a] chiropractic treatment with [b] chiropractic treatment plus SAP; for the results to have had any meaning, a control group involving sham SAP treatment would have been a minimal requirement. The authors' conclusions as published are therefore highly misleading.

In the discussion section of this paper, the authors admit that "*This study lacked a true control group*", claiming that that was because "*it was not ethically possible within the context of the study's clinical facility to include such a group*", and expressing that larger studies in the future should indeed include a control group. We find it regrettable that despite this frank admission by the paper's authors on the fundamental issue of lack of a true control group, the journal was not deterred from publishing this paper, and we would argue that the inclusion of a 'true control group' was certainly not ethically impossible, rather it should be considered ethically mandatory.

---

[4]The threshold was a reduction in headache frequency greater than 40%.

1.6   No Placebo Control Group: Aromatherapy, Reflexology and Rheumatoid Arthritis

The above study did not include an adequate control group, but the authors were open about it and readers could therefore draw their own conclusions (provided they read more than just the study's summary). Unfortunately, many CAM studies include a comparator group which, despite being described by the researchers as a control group, in fact does not constitute a true control and therefore only serves to confuse the reader. A common manifestation of fundamental error (or deceit?) is to have a 'control' that does not include a placebo: in other words, the 'control' subjects are simply untreated patients. Thus, a control group appears to exist, and can be referred to as such by the authors; yet this group provides such a poor comparator that studies of this type are rendered useless. The following paper, examining CAM therapies for rheumatoid arthritis, provides a typical example of this diversion tactic.

A total of 51 patients with rheumatoid arthritis were randomly assigned to one of three groups: (1) aromatherapy massage; (2) reflexology; or (3) a 'control' group (Gok Metin and Ozdemir 2016). Aromatherapy massage was applied to the knees of subjects in group (1) three times per week, and reflexology was administered to the feet of subjects in group (2) once per week. The treatments were delivered over a 6-week period. The subjects in group (3), the so-called 'control' group, received no intervention at all. Pain and fatigue scores were measured at baseline and within an hour after each intervention for the 6 weeks. These scores significantly decreased in groups (1 and 2) compared with group (3). The authors concluded that *"aromatherapy massage and reflexology are simple and effective non-pharmacologic nursing interventions that can be used to help manage pain and fatigue in patients with rheumatoid arthritis."*

So, the two interventions generated better outcomes than no therapy at all. Yet it is quite simply wrong to assume that this outcome is related to any specific effects of the aromatherapy or reflexology. Both treatments are agreeable; they involve touch, attention and care; and they generate expectations. These factors are very likely to produce positive outcomes. Thus, the effects reported in this study could be entirely unrelated to any specific therapeutic actions of aromatherapy or reflexology.

Therefore, the 'positive' results of this study might be due, at least in part, to the placebo effect. The only way to discern between placebo effect and any specific therapeutic benefits from these CAM treatments would have been to include a placebo control group in the experimental design. This group should have comprised patients treated with a placebo version of the treatment under investigation. In the case of the present study, this should have entailed a group receiving sham aromatherapy massage and another receiving sham reflexology, with both sham procedures applied in the same setting and with the same duration and frequency as the real therapies.

Unfortunately, this is just one example of a plethora of studies that claim to have a control arm which, on closer inspection, turns out to be unworthy of that description because no placebo has been used. Such studies almost inevitably present misleading conclusions. With flawed studies of this sort polluting the literature of CAM to the extent that they currently do, the public is going to be systematically misinformed. The consequence is that wrong therapeutic decisions are unavoidable.

1.7 No Placebo Control Group: CBT, Reiki and Depression

For our next example, we return to Reiki. This RCT intended to investigate the effectiveness of Reiki as a treatment for depression (Charkhandeh et al. 2016). In our earlier example involving this therapy (as a form of stress management), there was no control group at all. By contrast, the present study purports to include a control group—but it is not a true control group. As with our last example, the control group is, in effect, a no-treatment group.

The main aim of this study was to investigate the effectiveness of Reiki in reducing depression scores in adolescents. Reiki was tested alongside cognitive behavioural therapy (CBT), a common conventional treatment for depression. The researchers recruited 188 depressed adolescents who were randomly assigned to Reiki, CBT, or a wait-list control group. As with the above example, this latter group was simply a placebo-free 'no treatment' group.

Depression scores were assessed before and after 12 weeks of treatment/wait list. CBT showed a significantly greater decrease in depression scores than both Reiki and the controls. Reiki also showed greater decreases in depression scores relative to no treatment. The authors concluded that "*both CBT and Reiki were effective in reducing the symptoms of depression over the treatment period, with effect for CBT greater than Reiki.*"

The conclusion that Reiki was "*effective in reducing the symptoms of depression*" is based on the comparison of Reiki with no treatment. Reiki involves expectation, time, attention and empathy, factors which can be assumed to generate a placebo effect—most probably a sizeable one, given the subjective and labile nature of depressive symptoms. Additionally, because a proportion of the patients in the no treatment control group must have been disappointed for not getting such attention, they can be assumed to have experienced the adverse effects of their disappointment. These two phenomena combined can easily explain the result without any "effectiveness" of Reiki per se. In other words, the conclusions of the paper are misleading.

It is notable that the paper was published in a respected medical journal, *Psychiatry Research.* Speculatively, it is possible that the journal published such a flawed paper because many of the features of a rigorous study were present, not least the apparent inclusion of a control group. Medical journals of decent calibre seldom publish studies that plainly omit a control group; in this regard, it is notable that the majority of 'no control' papers are published in either specialist CAM journals—with the low standards that prevail there—or in low-ranking medical/scientific journals, which often have difficulties in attracting suitable reviewers (or good quality papers).

We consider that there is an ethical onus residing with all journal editors and reviewers to consider the nature of claimed 'control' groups in sufficient depth as to be able to discern the real thing from its various illusory forms. The reality is that this deeper scrutiny frequently does not occur, as evidenced by the large number of studies published that claim to have used a control arm, but which nevertheless turn out to have used merely a facade of such a feature.

1.8   Pseudo-control Group: CAM and Cancer Chemotherapy

The inclusion of an invalid comparator group looks better than having no comparator group whatsoever—at least at first sight. The last two examples show one means of achieving this semblance of rigour, namely by having a CAM group versus a no-treatment 'control' group. But there are other ways to make a study appear as if it has a valid comparator group, while in fact avoiding a true control group. One such subterfuge is to assign participants who chose not to receive the full course of CAM treatment to a 'non-comply' group, and use the data from this group as a comparator for the results of the patients who received the CAM treatment.

This type of study design is beautiful in its simplicity: one merely administers a treatment or treatment package to a group of patients; inevitably some patients comply, while others don't. The fact that some patients do not want the treatment provides the researcher with two groups of patients: those who do and those who do not comply. The investigator can now make the non-compliers appear like a proper control group and compare the outcomes of the two groups; we call this a 'pseudo-control' group. This trial design will make almost any treatment look efficacious, even one that is a mere placebo.

To illustrate this, we shall consider a trial with a promising title: *Quality-of-life outcomes in patients with gynaecologic cancer referred to integrative oncology treatment during chemotherapy* (Ben-Arye et al. 2015). It involved 128 patients who were referred to an 'integrative physician' (IP) for a consultation and subsequent treatment with a wide range of CAM therapies.[5] Patients were evaluated at

---

[5]The paper states that therapeutic interventions offered by the IP consultation included: "*herbal and dietary supplements and weekly acupuncture sessions, often in combination with mind-body (relaxation techniques, guided imagery, music therapy, etc.) or manual (e.g., acupressure) techniques.*"

the initial IP consultation and at 6–12 weeks thereafter, using questionnaire tools designed to assess symptoms and wellbeing. Of the 128 patients referred, 102 attended the initial IP consultation. After the trial, 96 of these patients (those who were still participating in the study) were assigned to two groups, 'AIC' (n = 68) and 'non-AIC' (n = 28). AIC refers to 'adherence to the integrative care'; patients were assigned to this group where they met a threshold for compliance.[6] The remainder of the patients were assigned to the non-AIC group.

At first sight, these findings appear impressive: patients' fatigue scores improved by on average 1.97 points (on a scale of 0–10) in the AIC group while they worsened by 0.27 points in the non-AIC group (a difference that was statistically significant). Similarly, patient concerns and wellbeing scores improved significantly in the AIC group but not in the non-AIC group. Certainly, the study's authors thought that their results were worthwhile, concluding: *"An IP-guided CM treatment regimen provided to patients with gynecological cancer during chemotherapy may reduce cancer-related fatigue and improve other QOL outcomes"*.

But this study is deeply flawed. The reasons for patients not complying could range from lack of perceived effectiveness to experience of side-effects. This means that the non-AIC group will comprise patients who will be fundamentally different from the compliant patients. Accordingly, the non-AIC group most certainly does not serve as a valid control group, and the data from this group cannot provide a valid comparison. Once again, lack of a true control group renders the results worthless.

1.9 Pseudo-control Group: Homeopathy and Respiratory Tract Infections

Studies like the ones cited above typically lack a control arm or use an inappropriate comparator group in lieu of a true control arm. In some cases, the authors do not explicitly claim to use a true control group—this is so in the previous example. Unfortunately, many published CAM studies claim to have a control arm which, on closer inspection, turns out to be worthless. The following study provides one such example.

This study was intended "to investigate the role of the homeopathic medicine Oscillococcinum® in preventing respiratory tract infections (RTIs)" (Beghi and Morselli-Labate 2016). It involved a retrospective analysis of patients' medical records and included 459 patients who had been referred to a respiratory diseases specialist. 248 patients were treated with the homeopathic medicine, while 211 were not treated. The latter group was deemed to be the control group. All patients were followed-up for at least one year, and up to a maximum of 10 years.

---

[6]The compliance threshold was ≥ 4 CM treatments, with ≤ 30 days between each session.

A significant reduction in the frequency of onset of RTIs was found in both the homeopathic medicine and the 'control' group. The reduction in the mean number of RTI episodes during the period of observation compared to the year before inclusion in the study was significantly greater in the homeopathic group than in untreated patients. The number of infections during the follow-up period is plotted in the following graph.

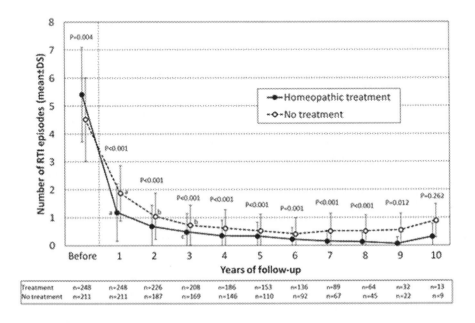

The authors concluded: "*these results suggest that homeopathic medicine may have a positive effect in preventing RTIs. However, randomized studies are needed before any firm conclusion can be reached*".

This study is flawed in several respects, most prominent of which is that the 'control group' is not, in fact, a true control. This group was not 'not treated' as the authors claim; it turns out that these patients were originally prescribed the homeopathic remedy but did not comply. This suggests an alternative and more plausible explanation for the findings, based on a systematic difference between the experimental and 'control' groups, as follows.

It is possible or even likely that the group of patients who did not comply would disproportionately comprise those who were more severely ill. This is so because the conditions of these patients would have been less susceptible to 'get better anyway' improvements or the placebo effect. Consequently, these patients would be less likely to experience a 'response' coinciding with their use of homeopathy. They would therefore be at high probability of dropping out of the study, as the homeopathy was evidently not working for them. (Some of these patients may have gone on to obtain conventional therapy for their RTIs; such treatment may have had positive effects, but the more severe nature of their condition would limit the improvements observed.)

So, in this study the 'control' patients differed systematically from the experimental group. It is therefore of no surprise to discover that they showed different clinical results. This alternative explanation is, of course, speculative, but it exemplifies the importance of comparing groups that are comparable, i.e. the importance of using a true control group.

Before passing on from this study, it is worth briefly noting some of its other flaws: [a] a follow-up between 1 and 10 years is an unacceptable variation for a research project of this nature; [b] before the start of the study, patients had more than 5 infections per year—this is way beyond the normal average of 1–2, and brings into question the representativeness of the volunteer patients; and [c] the primary outcome—the reduction in the average number of RTI episodes per year—was assessed by simply asking the patients, a method which is wide open to recall-bias and therefore not suited as a primary outcome measure in a clinical trial.

We conclude that, contra the claims of its authors, this study offers no reliable evidence whatsoever that homeopathic medicine may have a positive effect in preventing RTIs. The authors do admit a 'number of limitations' with their study, stating that "*the use of an untreated control group, rather than a placebo-treated one, might be questionable*". Apart from the fact that this group was probably not untreated (see above), we consider that the concept of an 'untreated control group'—which was central to this study—is not merely 'questionable' but so flawed that the results from this study and others like it testing therapeutic effectiveness are rendered valueless. From an ethical perspective, this approach is unacceptable because, once again, it is bound to generate misleading conclusions as to the supposed effectiveness of CAM.

2 'A + B Versus B' Trials

Another popular means of avoiding an adequate control group is the use of a design that we shall refer to as 'A + B versus B' (Ernst 2012). In such a study, patients are randomized to receive either: [i] CAM treatment (A) together with usual care (B); or [ii] usual care (B) alone. This looks rigorous and can be published as a 'pragmatic' trial addressing a 'real-life' problem. Moreover, because A + B is always more than B alone, even if A is a pure placebo, this design has the amazing advantage of never failing to produce a positive result! A few examples of recently published trials illuminate this fact.

2.1 'A + B Versus B' Design: Manipulation, Exercise and Leg Pain

Our first example is a trial involving chiropractic therapy and home exercise for leg pain (Bronfort et al. 2014). According to the authors, their study was '*to determine whether spinal manipulative therapy (SMT) plus home exercise and advice (HEA) compared with HEA alone reduces leg pain in the short and long term in adults with sub-acute and chronic back-related leg-pain (BRLP).*' Patients with back-related leg pain (BRLP) for least 4 weeks were randomised to receive either: [i] 12 weeks of SMT plus HEA; or [ii] HEA alone. Eleven chiropractors with a minimum of 5 years of practice experience delivered SMT in both groups.

The primary outcome was subjective BRLP at 12 and 52 weeks. Six secondary outcomes (at 12 and 52 weeks) were included: self-reported low back pain, disability, global improvement, satisfaction, medication use, and general health status. Of the 192 enrolled patients, 191 (99%) provided follow-up data at 12 weeks and 179 (93%) at 52 weeks. For leg pain, SMT plus HEA had a clinically important advantage over HEA at 12 weeks but not at 52 weeks. Nearly all secondary outcomes improved more with SMT plus HEA at 12 weeks, but only 3 (from 6) showed sustained improvements at 52 weeks. The authors concluded: "*for patients with BRLP, SMT plus HEA was more effective than HEA alone after 12 weeks, but the benefit was sustained only for some secondary outcomes at 52 weeks.*"

As this study design can hardly ever generate a negative result, it is not a true test of the experimental treatment. Rather, it can be viewed as a reliable means to create a false-positive finding for a potentially useless treatment. Had the investigators tested any other mildly pleasant placebo involving attention, time and touch, the result would have been the same. The conclusion that the SMT generates specific effects, as is implicit in this article, is unwarranted and potentially misleading. A more accurate conclusion could have been as follows: *SMT plus HEA was associated with better outcomes than HEA alone after 12 weeks, but the benefit was sustained only for some secondary outcomes at 52 weeks. Because the trial design did not control for non-specific effects, the observed outcomes are consistent with SMT being a placebo.*

Research where the result is known before the study has even started (i.e. any study with the 'A + B vs. B' design) is not just useless; it is, in our view, unethical: it fails to answer a meaningful question and is a waste of resources as well as an abuse of patients' willingness to participate in clinical trials.

2.2 'A + B Versus B' Design: Acupuncture and Hot Flashes

The flawed 'A + B versus B' design can be used with any CAM modality. The authors of one study wanted to determine the effectiveness of acupuncture in the management of hot flashes in women will breast cancer (Lesi et al. 2016). This paper looks impressive: it was published in a high-ranking medical journal and describes a large, randomized clinical trial. A total of 190 women with breast cancer were randomly assigned to two groups. Both groups received a booklet with information about climacteric syndrome and its management. In addition, the acupuncture group received 10 traditional acupuncture treatment sessions involving needling of predefined acupoints. The primary outcome was hot flash score after the 12 weeks of therapy, and secondary outcomes were climacteric symptoms and quality of life scores. Health outcomes were measured for up to 6 months after treatment.

At the end of treatment and at post-treatment follow-up, acupuncture plus enhanced self-care ('A + B') was associated with a significantly lower hot flash score than enhanced self-care alone ('B'). The 'A + B' treatment was also associated with fewer climacteric symptoms and higher quality of life scores. The authors concluded that *'acupuncture in association with enhanced self-care is an effective integrative intervention for managing hot flashes and improving quality of life in women with breast cancer.'*

Just as with the above chiropractic trial, these results provide no evidence of clinical improvement beyond placebo. The fact that the respected *Journal of Clinical Oncology* saw fit to publish this paper is regrettable, and demonstrates the unfortunate reality that, for reasons discussed above, publication of questionable CAM research is not restricted to low-ranking or sectarian journals.

2.3 'A + B Versus B' Design: Mindfulness, CRT and Back Pain

The power of the 'A + B versus B' design to create the impression of a robust study and provide impressive-looking results is further evidenced by our next example. This is an RCT aimed at evaluating treatments for chronic low back pain (Cherkin et al. 2016). At first glance, this seems like a sound study. It was conducted by one of the leading US back pain research team and was published in *JAMA*, a top medical journal. However, on closer examination, serious doubts emerge about several aspects of this trial.

The study considered two specific treatments: mindfulness-based stress reduction (MBSR) and cognitive behavioural therapy (CBT). MBSR meant training in mindfulness meditation and yoga, and is a form of CAM—albeit at the less 'extreme' end of the CAM spectrum, in that there is some plausibility in the notion that forms of relaxation may alleviate pain symptoms. CBT meant training to change pain-related thoughts and behaviours. CBT is commonly used to treat psychological or psychiatric conditions such as anxiety and depression, in which context it is a mainstream therapy. Its use for pain control, however, is less usual and arguably amounts to a form of CAM. Semantics aside, the important question about these therapies is whether they can reduce pain.

The investigators randomly assigned 342 chronic back pain patients into three groups receiving either MBSR, CBT or 'usual care'. Both MBSR and CBT were delivered in 8 weekly 2-h sessions. Usual care meant whatever care participants received outside of the trial. Clinical assessment was conducted at various junctures up to 52 weeks. The primary outcomes were the percentages of participants with clinically meaningful improvement in two parameters: [a] functional limitations and [b] self-reported back pain bothersomeness. Here are the key results at 26 weeks:

- Functional improvement was higher in the group who received MBSR (60.5%) and CBT (57.7%) than for the usual care group (44.1%)
- Improvement in pain bothersomeness was higher in the MBSR group (43.6%) and the CBT group (44.9%), compared with the usual care group (26.6%)

The authors concluded that "*among adults with chronic low back pain, treatment with MBSR or CBT, compared with usual care, resulted in greater improvement in back pain and functional limitations at 26 weeks, with no significant differences in outcomes between MBSR and CBT. These findings suggest that MBSR may be an effective treatment option for patients with chronic low back pain.*"

The authors state that they aimed at evaluating "*the effectiveness for chronic low back pain of MBSR vs. cognitive behavioural therapy (CBT) or usual care.*" This is not just misleading, it is incorrect! The method section includes the following crucial and revelatory statement: "*All participants received any medical care they would normally receive.*" Thus, the correct stated aim should have been to evaluate *the effectiveness for chronic low back pain of MBSR plus usual care versus cognitive behavioural therapy plus usual care or usual care alone.* Consequently, the author's conclusions are equally wrong. They should have read as follows: *Among adults with chronic low back pain, treatment with MBSR plus usual care or CBT plus usual care, compared with usual care alone, resulted in greater improvement in back pain and functional limitations at 26 weeks, with no significant differences in outcomes between MBSR and CBT.*

In other words, this is yet another trial with the 'A + B versus B' design. As with our other examples, because A + B is always more than B (even if A is just a placebo), this study design could never have generated a negative result! The results are therefore entirely compatible with the notion that the two tested treatments, MSBR and CBT, are pure placebos. Add to this the disappointment many patients in the 'usual care group' might have felt for not receiving an additional therapy for their pain, and you have a most plausible explanation for the observed outcomes.

The 'A + B versus B' design can only produce positive findings. Any such study allegedly testing the effectiveness of therapy XY and concluding that 'it is effective' ought to be categorised as unethical pseudo-science.

3 Problems with Results from CAM Studies

We shall now set out some examples of misleading analysis of the data from CAM studies. Our examples deal with the use of secondary endpoints, statistical mal-practice, and results that raise eyebrows because they are simply too good to be true. Again, we ought to stress that these are mere examples of a vast number of such ethical breaches.

3.1 Secondary Endpoints: Craniosacral Therapy and Back Pain

A classic way to produce misleading research findings is the non-reporting of parts of the data that turned out to disagree with the desired conclusion. As discussed in the last chapter, in most studies, researchers include a large number of measured parameters. Once the statistics of a trial have been calculated, it is likely that some of them yield positive results purely by chance—in statistical language, this is a multiple comparisons effect. By simply omitting any mention of the negative results obtained for the primary endpoint, a researcher can easily turn a fundamentally negative study into a seemingly positive one. Normally, scientists must rely on a pre-specified protocol which defines a primary outcome measure. However, in the absence of proper governance, it is possible to publish a report which obscures such detail and thus misleads the public.

An example is provided by the following RCT evaluating the effects of cran-iosacral therapy (CST) in patients with low back pain (Castro-Sanchez et al. 2016). CST involves light touch applied mainly to the cranium. CST proponents believe that this creates subtle movements at the sutures (joints) between the bony plates making up the skull which, in turn, correct fluid and energy imbalances within the body with concomitant therapeutic benefits. These assumptions are sheer fantasy: there is no physiological or anatomic evidence for significant cranial plate 'movements'; nor is there any basis for the notion that 'fluid imbalances' exist, cause disease, or can be altered by touch (Ernst 2015).

In this study, the researchers wanted to investigate the effects of CST on dis-ability, pain intensity, quality of life, and mobility. They assigned 64 patients with chronic non-specific low back pain to either [a] the experimental group, receiving 10 sessions of CST; or [b] the control group, receiving 10 sessions of classic massage. (Craniosacral therapy took 50 min and was conducted per a pre-specified protocol. The classic massage protocol took 30 min.) This design is better than most of the examples described above, in that a placebo control group was used, with the classic massage serving in effect as a sham form of treatment. However, substantial flaws remained: the patients were not blinded as to which therapy they received, and the control group received a total of 200 min less attention than the CST group.

Disability, as measured by a questionnaire-based approach, was the primary endpoint. A myriad of additional parameters was also assessed, including: pain intensity; kinesiophobia (fear of moving); endurance of trunk muscles; lumbar mobility; haemoglobin oxygen saturation; systolic blood pressure; diastolic blood pressure; blood flow parameters; and biochemical analyses of fluids. All outcomes were measured at baseline, after treatment, and at one-month follow-up.

For the primary outcome measure—disability—the results showed no statistically significant differences between the CST and control massage groups. In other words, the results of this RCT were essentially negative. However, patients receiving CST experienced greater improvement in a few of the many secondary endpoints, namely: pain intensity; haemoglobin oxygen saturation; systolic blood pressure; serum potassium; and serum magnesium.

The authors concluded that "*10 sessions of craniosacral therapy resulted in a statistically greater improvement in pain intensity, hemoglobin oxygen saturation, systolic blood pressure, serum potassium, and magnesium level than did 10 sessions of classic massage in patients with low back pain.*"

This conclusion is misleading. The reporting of and emphasis on secondary endpoints is a classic example of a phenomenon we referred to earlier in this chapter as 'data-dredging': sift through a large quantity of data and there is a good chance of finding some apparently 'significant' occurrences that are simply statistical artefacts. Unwitting or unscrupulous researchers can then suggest that such chance findings are evidence of a therapeutic effect.

It is clear that this paper provides no good evidence that the small number of 'positive' outcomes were, in fact, caused by the CST.[7] This publication is an example of an attempt to turn a negative result into a positive one, a phenomenon which is embarrassingly frequent in CAM. Clearly, this is not just misleading but also dishonest, unethical and potentially harmful to patients.

3.2   Statistical Malpractice: Homeopathy and Postoperative Recovery

We already have highlighted the pitfalls of using $p = 0.05$ as a criterion of statistical significance. We noted that some contemporary statisticians recommend that, short of adopting a Bayesian statistical approach, the more stringent threshold of $p = 0.005$ ought to be employed. Our criticisms of $p = 0.05$ apply to the majority of CAM

---

[7] An alternative explanation for the (few) positive results of this study, other than these results merely being statistical artefacts, is that the observed improvements could have been a result of various flaws in the study design: in particular, the lack of patient-blinding, and the CST group receiving 200 min longer attention than the control patients. Distinguishing between these two explanations is impossible from the data generated by this study. In either case, our judgement of this paper remains unaltered: its data are essentially meaningless, and the authors have used them to present misleading conclusions.

studies, including most of the examples given in this chapter. However, when even the $p = 0.05$ threshold does not yield the desired positive result, some CAM researchers simply move the goal-post.

The following study of homeopathic arnica for post-operative facial bruising (*ecchymosis*) serves as an example for this phenomenon (Chaiet and Marcus 2016). Subjects scheduled for nasal surgery (rhinoplasty) entailing removal of bone (osteotomy) were randomized to receive either oral arnica or placebo in a double-blinded fashion. A commercially available arnica preparation was used, immediately before and two days after surgery. The extent and intensity of bruising were measured over the next few days, using digital analysis of photographs of the patient's face.

Compared with 13 subjects receiving placebo, 9 taking arnica had 16.2, 32.9, and 20.4% less extensive bruising on postoperative days 2/3, 7, and 9/10. In terms of intensity of bruising, there was initially a 13.1% increase in intensity with arnica, but 10.9 and 36.3% decreases on days 7 and 9/10.

The researchers tested their data statistically, and reported two 'statistically significant' results: reduction of bruising extent on day 7, and reduction of bruising intensity on day 9/10. (All the other changes did not pass their test of significance.) They concluded: "*Arnica montana seems to accelerate postoperative healing, with quicker resolution of the extent and the intensity of ecchymosis after osteotomies in rhinoplasty surgery, which may dramatically affect patient satisfaction.*"

Considering [a] that homeopathy is implausible, and [b] that previous systematic reviews confirm that homeopathic arnica is a pure placebo (Ludtke and Hacke 2005), these results seem surprising. But the puzzle can be solved easily: if we apply the conventional level of statistical significance, $p = 0.05$, there are no statistically significant differences to placebo at all! One must dig deep into the methods section of the article to find the following sentence: "*a P value of 0.1 was set as a meaningful difference with statistical significance.*" So, the authors moved the goal-post. In fact, none of the effects called 'significant' by the authors pass the conventionally used probability level of 5%—even though, as discussed above, this level itself is easily reached, especially in weak trails, and is associated with an extremely high false discovery rate. In other words, contrary to what the authors try to make us believe, the results show that homeopathic arnica has no specific effects.

3.3 Too Good to Be True? Bach Flower Remedies and Carpal Tunnel Syndrome

The most direct method of misguiding people with research findings is, of course, to simply cheat. Today there are well over 100 CAM journals competing for the very few high-quality papers that emerge in this lacklustre field. Consequently, many journals publish articles which are seriously flawed. As discussed earlier, the process of peer-review is a mechanism supposed to minimise the risk of scientific errors and fraud. Yet the journals in question tend to have a peer-review process that rarely involves independent and critical scientists; often authors can even ask for their friends to do the peer-review, and the journal will follow that recommendation. Thus, the door is wide open to cheating.

It is rarely possible to directly identify a fraudulent paper once it is published. However, sometimes one gets the impression that the findings seem too good to be true. Here is one such example.

A randomized, placebo-controlled clinical trial was conducted with the aim of evaluating the effectiveness of a cream based on Bach Flower Remedies (BFR) for symptoms of carpal tunnel syndrome (Rivas-Suarez et al. 2017). Forty-three patients with mild to moderate carpal tunnel syndrome during their "waiting" time for surgery were randomized into 3 parallel groups: Placebo (n = 14), blinded BFR (n = 16), and non-blinded BFR (n = 13). These groups were treated during 21 days with topical placebo or a cream based on BFR.

The reported results were impressive. Significant improvements were observed on self-reported symptom severity and pain intensity favourable to the two BFR groups. In addition, all signs observed during the clinical exam showed significant improvements among these groups as well as symptoms of pain, night pain, and tingling. Statistical analysis revealed the data to be robust: the effect sizes of the clinical improvements were large, and the p-values were low, with several key improvements being significant at $p < 0.001$. The authors of this study concluded: "*the proposed BFR cream could be an effective intervention in the management of mild and moderate carpal tunnel syndrome, reducing the severity symptoms and providing pain relief.*"

Bach flower therapy is based on the notion that wildflower extracts, highly diluted with brandy and water, can combat a lack of 'harmony' in the patient and thus cure disease (Carroll 2017c). Like homeopathic remedies, BFR preparations contain no active molecules. It seems therefore most implausible that such a tiny study of BFR might generate such large and statistically robust effects.

We can, of course, not say for certain that cheating was involved in this case. An alternative explanation is that the results were simply due to chance alone. However, the multiple low p-values and high effect sizes make this explanation one of low likelihood.

3.4   Too Good to Be True? Wet Cupping and Back Pain

Another example of a study that looks too good to be true is a clinical trial (Al Bedah et al. 2015) evaluating the effectiveness and safety of wet cupping therapy for persistent non-specific low back pain (PNSLBP). Wet cupping involves puncturing the skin and placing an inverted cup over the injured area. A vacuum is created in the cup, which is believed by practitioners to allow 'toxins' to be sucked out of the body, thus restoring health (Carroll 2017b).

The investigators recruited 80 patients with PNSLBP that had lasted at least 3 months and randomly allocated them to an intervention group ($n = 40$) or to a control group ($n = 40$). The intervention group had 6 wet cupping sessions within 2 weeks. The control group had no such treatments. Acetaminophen (paracetamol) was allowed as a rescue treatment in both groups. In terms of outcome measures, pain and disability were assessed using questionnaires, and numbers of acetaminophen tablets taken were compared at 4 weeks from baseline.

At the end of the intervention, statistically significant differences in the outcome measures favouring the wet cupping group compared with the control group were seen ($p = 0.0001$). These improvements continued for another two weeks after the end of the intervention. The authors concluded: "*wet cupping is potentially effective in reducing pain and improving disability associated with PNSLBP at least for 2 weeks after the end of the wet cupping period. Placebo-controlled trials are needed.*"

Aside from numerous weaknesses of the study design (prominent amongst which is the lack of a placebo group), the results per se are simply not plausible. Low back pain has a natural history that is well-studied. We therefore know that most cases do get better within 2–3 weeks regardless of whether we treat them or not. In this study, the control group did not improve at all.

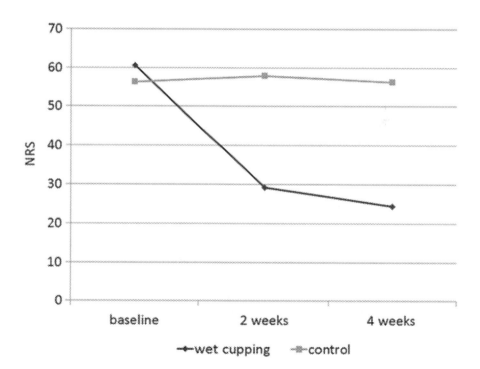

The improvement of the experimental group looks much like the one might expect from the natural history of back pain. If this were true, the effect of wet cupping would by close to zero and the conclusion drawn by the authors of this trial would be false-positive. But why was there no improvement in the control group? We do not know the answer to this question. All we do know is that it is this unexplained phenomenon which has created the difference between the groups and the impression of wet cupping being efficacious.

4   Fraud

In clinical research, outright cheating is today difficult. Investigators cannot easily claim to have conducted a clinical trial without a complex infrastructure which invariably involves other people who are likely to want to have some control. This means that complete fabrication of an entire data set is probably a rare event. What is more feasible, however, is the 'prettification' of the results. By just 'readjusting' a few data points that failed to live up to the investigator's expectations, a researcher might be able to turn a negative into a positive trial. Proper governance is aimed at preventing his type of 'mini-fraud' but, in CAM, such mechanisms are rarely adequately implemented. Even in mainstream medical research, such deceit is very difficult to detect, and undoubtedly occurs.

Nevertheless, although it is less common than merely 'tweaking' the data, outright fraud does occur in medical research. In CAM research, this is a particular concern, given the lack of oversight of CAM trials, not to mention the possibility that CAM researchers may be incentivised to cheat due to the reality that their favoured therapy can have no specific effects. The following four cases provide instances of outright fraud an even criminal behaviour in CAM research, involving a wide range of CAM modalities.

4.1   Intercessory Prayer and Pregnancy

Amazingly, a 2001 study of the effects of intercessory prayer on pregnancy rates in women being treated with in vitro fertilization (IVF) seemed to show that such religious intervention is effective in improving IVF success rates (Cha and Wirth 2001). When the senior author, D. P. Wirth from the 'Healing Sciences Research International' in Orinda, California, went to prison in 2004 after being convicted of multiple cases of fraud, this paper, together with several further of his studies of paranormal healing (all showing positive results!), was withdrawn (Flamm 2004). Yet these studies continue to be cited by healing enthusiasts as proof of effectiveness of spiritual healing methods, and 8 papers of D. P. Wirth are still available on Medline today.

4.2 Evening Primrose Oil and Eczema

Most people would doubtless regard therapeutic intercessory prayer as being completely beyond the pale, and therefore would be unsurprised to learn that claims for its effectiveness are based on dishonest research. However, popular CAM therapies of the sort that appear more 'plausible' to the uninitiated have also been the subject of fraud. And such fraud sometimes has a lasting effect in terms of our belief in a therapy, as the following example illustrates.

One of the best-selling supplements is evening primrose oil (EPO); it is used for a range of conditions, including eczema. However, a systematic review of this topic drew clear conclusions: "*evening primrose oil lack(s) effect on eczema; improvement was similar to respective placebos used in trials ...EPO (is) not effective treatment for eczema*" (Bamford et al. 2013).

So, where does the very widespread notion that EPO is effective for eczema come from? It was originally promoted by David Horrobin, a researcher turned entrepreneur who claimed that several human diseases, including eczema, were due to a lack of fatty acid precursors and could thus be effectively treated with EPO. In the 1980s, Horrobin began to sell EPO supplements without having conclusively demonstrated their safety and effectiveness; this led to confiscations and felony indictments in the US. As chief executive of Scotia Pharmaceuticals, Horrobin obtained licences for several EPO-preparations which later were withdrawn for lack of effectiveness. Charges of mismanagement and fraud led to Horrobin being ousted as CEO by the board of the company.

Later, Horrobin published a positive systematic review of EPO for eczema where he excluded the negative results of the largest published trial, but included seven of his own unpublished studies. When scientists asked to examine the data, Horrobin's legal team convinced the journal editor to refuse the request. In fact, the evidence for EPO is negative not just for eczema. There is not a single disease or symptom for which it demonstrably works. One review of the data concluded: "*EPO has not been established as an effective treatment for any condition*" (Ernst et al. 2006).

Horrobin misled all of us: patients, health care professionals, scientists, journal editors, regulators, decision makers, businesspeople. This caused unnecessary expense and set back research efforts in a multitude of areas. And EPO continues to the present day to be believed in by many consumers and even some health professionals. So, this instance of fraud has helped establish a major market in an unproven CAM therapy—a market that continues to thrive despite the thorough refutation of Horrobin's claims.

4.3 Homeopathy Test System

In 2003, two German pharmacists, Prof. Dr. Karen Nieber and Prof. Dr. Wolfgang
Süß of the University Leipzig received the Hans-Heinrich-Reckeweg-Preis, an
award given for those who have earned a reputation for 'homotoxicology'—a CAM
modality similar to homeopathy—for a paper entitled *Development of an in vitro
test system for proving the effects of selected homeopathic dilutions* (Adams 2017).[8]
Their research seemed to show that it was not the dilution but the succussion
(vigorous shaking, with apparently magical powers) that made homeopathic
preparations effective. When 'reasonable doubts' emerged on the reliability of their
findings, Prof Nieber returned the award, but Prof Suess did not. In 2005, he
received another homeopathy award, the Hahnemann-Preis. And as with the previous example, the negative effects of this fraud live on: the results of Suess and
Nieber continue to be cited as a proof of homeopathy's plausibility.

4.4 Ukrain and Cancer

We briefly mentioned the CAM cancer 'cure' known as Ukrain in Chapter 1.
Ukrain is based on alkaloids from the greater celandine plant and the compound
thiotepa. It was developed by Dr. Wassil Nowicky who allegedly cured his
brother's testicular cancer with his invention. Despite its high cost of about £50 per
injection, Ukrain has become popular in the UK and elsewhere. Ukrain has its name
from the fact that the brothers Nowicky originate from the Ukraine, where also
much of the research on this drug was conducted.

   In 2005, one of the present authors conducted a review of all the clinical studies
which had tested the effectiveness of Ukrain (Ernst and Schmidt 2005). Seven
randomised clinical trials were found, all of which reported baffling cure rates but
were methodologically weak. And these studies were odd in several other ways.
Their results seemed too good to be true. The authors of the studies seemed to
overlap and often included Nowicky himself. They were published in only two
different journals of low impact. The only non-Ukrainian trial came from Germany
and was not much better: its lead author happened to be the editor of the journal
where it was published. More importantly, the paper had a very small sample size
and lacked crucial methodological details, which rendered the findings difficult to
interpret.

   Collectively, these circumstances prompted our cautious conclusion that "numerous caveats prevent a positive conclusion". Sadly, our caution was not heeded
and many cancer centres around the world began to take Ukrain seriously; several
integrative cancer clinics even started using the drug in their clinical routine. Soon,
numerous websites sprang up praising Ukrain: "*It is the first medicament in the*

---

[8]This paper is in German; its title has been translated for this book.

world that accumulates in the cores of cancer cells very quickly after adminis-tration and kills only cancer cells while leaving healthy cells undamaged"; Ukrain's inventor and patent holder Dr. Wassil Nowicky was "*reported to have been nominated for the Nobel Prize in 2005*" (Nordfors 2017).

Somehow, we doubt this notion about the Nobel Prize. What we do not question, however, is this press release by the Austrian police: "*the Viennese police have been investigating Dr. Nowicky. During a 'major raid' on 4 September 2012, he and his accomplices were arrested under the suspicion of commercial fraud. Nowicky was accused of illegally producing and selling the unlicensed drug Ukrain. The financial damage was estimated to be in the region of 5 million Euros.*" (Hejl 2012).

5 Systematic Reviews

As we have seen, for a range of reasons, a single clinical trial can produce mis-leading or false findings. The only fair solution is to consider the totality of the available evidence and, crucially, to account for the reliability of each the studies published. As we outlined in the previous chapter, research projects that do so are called systematic reviews. This approach can produce the most reliable information possible as to the clinical effectiveness of a given therapy. However, some sys-tematic reviews instead present erroneous and misleading conclusions. Regrettably, systematic reviews of CAM therapies are frequently of the latter variety. Below are three examples of flawed systematic reviews in CAM.

5.1 *Acupuncture and Cognitive Impairment*

The aim of this systematic review was to "*estimate the clinical effectiveness and safety of acupuncture for amnestic mild cognitive impairment (AMCI)*", AMCI being the transitional stage between the normal memory loss of aging and dementia (Deng and Wang 2016). The review identified all RCTs of acupuncture versus medical treatment for AMCI. Five RCTs involving a total of 568 subjects were included. The methodological quality of the RCTs was found to be generally poor. Participants receiving acupuncture had better outcomes than those receiving the drug nimodipine, a result that was statistically significant ($p < 0.01$). Acupuncture used in conjunction with nimodipine significantly improved outcomes compared to nimodipine alone. Three trials reported adverse events.

The authors concluded: "*acupuncture appears effective for AMCI when used as an alternative or adjunctive treatment; however, caution must be exercised given the low methodological quality of included trials. Further, more rigorously designed studies are needed.*"

The authors try to tell us that their aim was "*to estimate the clinical ... safety of acupuncture...*" However, it is never possible to estimate its safety based on just a few clinical trials. To assess therapeutic safety, one would need sample sizes that go two or three dimensions beyond those of RCTs. Thus, safety assessments are best done by evaluating the evidence from *all* the available evidence, including case-reports, epidemiological investigations and observational studies. In other words, the aim of this systematic review, as formulated by its authors, is misguided.

Furthermore, the authors tell us that "two studies did not report whether any adverse events or side effects had occurred in the experimental or control groups." This deficit is a common and serious flaw of many trials of CAM therapies, and another important reason why RCTs cannot be used for evaluating their risks. Too many CAM studies simply don't mention adverse effects at all. If such studies are subsequently submitted to systematic reviews, they must generate a false positive picture about the safety of acupuncture. The absence of adverse effects reporting is a serious breach of research ethics. But in the realm of CAM, it is extremely common.

The authors conclude that acupuncture is more effective than nimodipine. This sounds impressive and might convince many readers—unless they happen to know that nimodipine is not supported by good evidence of effectiveness. A systematic review—this time of the robust variety—provided "*no convincing evidence that nimodipine is a useful treatment for the symptoms of dementia, either unclassified or according to the major subtypes – Alzheimer's disease, vascular, or mixed Alzheimer's and vascular dementia*" (Lopez-Arrieta and Birks 2000).

The authors also conclude that acupuncture used in conjunction with nimodipine is better than nimodipine alone. This too might sound impressive—unless one realises that all the RCTs in question failed to control for the effects of placebo and the added attention given to the patients (see our comments about the 'A + B vs. B' trial design earlier in this chapter). This means that the findings reported here are consistent with acupuncture itself being totally devoid of therapeutic effects.

The authors mention the paucity of RCTs and their mostly poor methodological quality. Yet they arrive at fairly definitive conclusions regarding the therapeutic value of acupuncture. However, based as they are upon a few poorly designed and poorly reported RCTs, even such tentatively positive conclusions are nonsensical and arguably unethical.

But the most serious flaw of this systematic review relates to the fact that all 5 RCTs that were included in the analyses were conducted in China by Chinese researchers and published in Chinese journals. It has been shown repeatedly that such studies hardly ever report anything other than positive results; no matter what conditions is being investigated, acupuncture turns out to be effective in the hands of Chinese researchers. This means that the results of such studies are known even before the first patient has been recruited. Little wonder then that virtually all reviews of such trials—and there are dozens of them—arrive at conclusions like those formulated in the paper before us.

And this precisely is the problem: systematic reviews by Chinese authors evaluating TCM therapies (which includes acupuncture) are arriving at misleading conclusions. Such papers are currently swamping the market. At first glance, they look fine. On closer scrutiny, however, most turn out to be stereotypically useless and promotional. The typical article starts by stating its objective which usually is to evaluate the evidence for a TCM as a treatment of a condition which few people in their right mind would treat with any form of TCM. It continues with details about the methodologies employed and then, in the results section, informs the reader that x studies were included in the review which mostly reported encouraging results but were wide open to bias. And then comes the crucial bit: the conclusions. These are as predictable as they are misleading.

A good example of such a conclusion is provided by another systematic review into Chinese herbal medicine to treat vascular dementia: "*This systematic review suggests that Chinese Herbal Medicine as an adjunctive therapy can improve cognitive impairment and enhance immediate response and quality of life in Senile Vascular Dementia patients. However, because of limitations of methodological quality in the included studies, further research of rigorous design is needed.*" (Zeng et al. 2015).

Another example of such a review, this one looking at TCM for venous ulcers, concluded: "*the evidence that external application of traditional Chinese medicine is an effective treatment for venous ulcers is encouraging, but not conclusive due to the low methodological quality of the RCTs. Therefore, more high-quality RCTs with larger sample sizes are required.*" (Li et al. 2015).

Such conclusions are highly problematic, for several reasons:

- We don't know what treatments the authors are talking about.
- Even if we tried to dig deeper, we cannot get the information because practically all the primary studies are published in obscure journals in Chinese language.
- Even if we did read Chinese, we probably would not feel motivated to assess the primary studies because we know they are all very poor quality—too flimsy to bother.
- Even if they were formally of good quality, we would have doubts about their reliability because 100% of these trials report positive findings!

Conclusions of this nature are deeply misleading and potentially harmful. They give the impression that there might be 'something in it', and that the treatment in question could be well worth trying. This may give false hope to patients and is therefore not ethical.

Sadly, the phenomenon of misleading systematic reviews is not confined to TCM; on the contrary, it is very common throughout the realm of CAM.

5.2 Eurythmy and Health

Eurythmy is a form of expressive movement which, as a purported therapy, is a component of the CAM modality known as anthroposophic medicine (Carroll

2017a). The aim of a 2015 systematic review was to update and summarize the relevant literature on the effectiveness of eurythmy in a therapeutic context since 2008 (Lotzke et al. 2015).

The 2015 paper is thus an update of a previously review, published in 2008 (Bussing et al. 2008). This previous review had found 8 trials which met the inclusion criterion. The methodological quality of these studies ranged from poor to good, and the sample size varied from 5 to 898 patients. In most cases, eurythmy was used not as a mono-therapy but as an add-on treatment. The studies reported positive effects with clinically relevant effect sizes in most cases. While acknowledging limitations with the studies, the authors concluded that eurythmy "*can be regarded as a potentially relevant add-on in a therapeutic concept, although its specific relevance remains to be clarified.*"

For the 2015 review, various databases were searched, and 11 publications met the inclusion criteria. These covered a wide range of study types, the methodological quality of which varied considerably. Most of the papers described positive treatment effects with varying effect sizes. The authors—who all come from the Institute of Integrative Medicine, anthroposophical University of Witten/Herdecke in Germany—drew the following conclusions: "*Eurythmy seems to be a beneficial add-on in a therapeutic context that can improve the health conditions of affected persons. More methodologically sound studies are needed to substantiate this positive impression.*"

The conclusions from both systematic reviews are truly bizarre. Each and every study reviewed by these authors is of insufficient quality to establish cause and effect. Both systematic reviews thus mislead the reader into thinking that the evidence for eurythmy is better than it truly is.

We believe that there is an important lesson here: the collation of individually weak or flawed papers—no matter how many such publications are included—does not amount to evidence that a therapy works. Garbage in, garbage out, they say. Yet, many CAM systematic reviews are based the logical fallacy that, if you add enough individually unbelievable reports together, the result is something believable.

CAM journals are particularly prone to publishing flawed systematic reviews (for instance, both above eurythmy reviews were published in CAM journals); but mainstream medical journals occasionally also publish highly questionable CAM reviews.[9]

---

[9]Recent (2017) examples include (Dong et al. 2017; Li et al. 2017; Ma et al. 2017; Shi et al. 2017; Sreenivasmurthy et al. 2017; Dai et al. 2017).

5.3 Naturopathy and Health

Our final example of flawed and misleading systematic reviews examined naturo-pathic medicine (naturopathy), an umbrella term which entails using only 'natural' remedies, such as sunlight, water, heat, cold, massage (Carroll 2017d). This review considered results from multi-modal treatment delivered by North American naturopathic doctors (Oberg et al. 2015).

Fifteen studies met the authors' inclusion criteria. They covered a wide range of chronic diseases. Effect sizes for the primary medical outcomes, including quality of life metrics, varied and were statistically significant in most studies. The authors concluded that "*previous reports about the lack of evidence or benefit of naturo-pathic medicine (NM) are inaccurate; a small but compelling body of research exists. Further investigation is warranted into the effectiveness of whole practice NM across a range of health conditions.*"

There are several important caveats here:

- The authors included all sorts of investigations, even uncontrolled studies; only 6 of the 15 investigations were RCTs.
- Rigorous trials were very scarce; and for each single condition even more so.
- The authors mention the PRISMA guidelines for systematic reviews implying that they followed them but, in fact, they did not.

Our biggest concern, however, relates to the interventions tested in these studies. All we have by way of an explanation is that the interventions tested in the studies included *diet counselling and nutritional recommendations, specific home exercises and physical activity recommendations, deep breathing techniques or other stress reduction strategies, dietary supplements including vitamins, hydrotherapy, soft-tissue manual techniques, electrical muscle stimulation, and botanical medicines.* This is a bewildering array of disparate therapies, of which the details of application are unknown. It is simply impossible to try to 'systematically review' such a vast catalogue of vague therapies in any coherent way that could yield believable findings. In sum, the review's conclusions about 'naturopathic medicine' are nonsensical, misleading and therefore unethical.

6 Surveys

We already pointed out that of all published CAM papers, a large proportion of those that contain empirical data are based on surveys of patients (as opposed to being clinical trials). We shall conclude our examination of the 'real world' of CAM research by looking at three examples from the hundreds if not thousands of such survey-based papers that have been published.

6.1   Homeopathy and Consumer Satisfaction

A good example of a typical CAM survey was presented at a recent conference of the American Public Health Association (APHA) by Alison Teitelbaum, a researcher from the US National Center for Homeopathy[10] (Teitelbaum 2017). It had an impressively large sample size, involving "*almost 20,000 consumers who had purchased at least 1 over-the-counter (OTC) homeopathic medicine in the past 2 years*". The stated justification for the survey was "*a marked increase in consumer demand for information*", coupled with the claim that "*very little data exists in the published literature*". The key results were very positive, as follows:[11]

- >95% of respondents indicated they were very or extremely satisfied with the most recent OTC homeopathic medicine they had purchased and used.
- >96% of respondents indicated they were very or extremely satisfied with the results of OTC homeopathic medicines that they had used in general.
- >98% of respondents reported that they were very likely to purchase OTC homeopathic medicines again in the future.
- >97% of respondents indicated that they were very likely to recommend homeopathic medicines to others.
- >80% of the respondents indicated using OTC homeopathic medicines for acute, self-limiting conditions, such as aches and pains; cold and flu symptoms; and digestive upset.

Teitelbaum concludes: "*These results support anecdotal evidence that homeopathic consumers are satisfied with OTC homeopathic medicines, and are using them to treat acute, self-limiting conditions. Additional research is needed to further explore the use of OTC homeopathic medicine in the US for trends, access, and overall awareness about homeopathy*".

Like many other CAM surveys, this paper contains numerous questionable assertions and dubious conclusions. The faults are evident right from the beginning. Take the claim that 'very little data exists': this statement is plain wrong; Medline has plenty of articles on this subject. In terms of the results, the precise number of respondents is not given; even worse, the response rate is not stated. These are very fundamental errors in survey based research—but sadly CAM surveys are frequently suffused with such basic blunders.

Turning to the respondents' views on OTC homeopathic medicines (as summarised in the first four bullet points above): the reported percentages look impressive, but one needs to bear in mind that only homeopathy fans were included in the survey! This investigation, and many others like it, is akin to doing a survey in a hamburger joint and concluding that all consumers love to eat hamburgers. We

---

[10]On their website, the NCH claims that they "inform legislators and work to secure homeopathy's place in the U.S health care system while working to ensure that homeopathy is accurately represented in the media" (NCH 2017).

[11]The numbers are ours.

do not know the views of other consumers who had bad experiences with OTC homeopathy medicines, including those who may have used them for anything more than the 'self-limiting' conditions reported by most respondents. In any case, the quoted percentages are effectively meaningless, since we do not know the response rate for the survey.

Further, the conclusion that 'additional research is needed' is a very common (indeed almost ubiquitous) invocation in CAM papers; we would strongly dispute this assertion and argue that more research would be a waste of resources, and would very likely lead to further misleading implications of therapeutic benefits from homeopathy. In other words, further research would be unethical!

One final point about such surveys: the author does not explicitly assert that homeopathy 'works'. However, this is implicit, and is bolstered by the fact that an academic journal has published it. Moreover, pro-CAM websites can 'marshal their armies' of publications referring to homeopathy (or whichever modality is being promoted), and present these on a long list, typically on a promotional website. While few consumers (or indeed health professionals) are likely to read the original papers, the impression is created that homeopathy is a *bona fide* subject for research. The implication conveyed to the unsuspecting reader of such propaganda is clear: 'it works'.

6.2 Natural Remedies and Tonsillopharyngitis

Our next example is a survey investigating the management of paediatric tonsillopharyngitis (sore throat possibly with symptoms such as fever), with a focus on 'natural remedies' (Salatino and Gray 2016). For that purpose, 138 paediatricians, general practitioners and ear-nose-throat (ENT) specialists from 7 countries were sent a non-validated questionnaire.

The results indicated that homeopathic remedies were prescribed as a supportive therapy by 62% of participants. In the chronic setting, homeopathy was suggested as a supportive therapy by 59% of all paediatricians, phytotherapy by 28%, and vitamins/nutritional supplementation by 37%.

The authors of this paper concluded from these results that "*the management of tonsillopharyngitis in paediatric patients still remains empiric. Natural remedies, and homeopathy in particular, are used in the management of URTIs. An integrative approach to these infections may help reduce excessive antibiotic prescription.*"

What could such a survey possibly show? That health care professionals who like homeopathy answer, while the majority don't? But the pinnacle of silliness must be the conclusions drawn from such 'research'. Let's take them step by step:

1. *The management of tonsillopharyngitis in paediatric patients still remains empiric*—'empiric' means that treatment is based on clinically educated guesses in the absence of sufficient research information; but this claim is simply not true.

2. *Natural remedies, and homeopathy in particular, are used in the management of URTIs*—this may be true, but it has been known before; we therefore do not need to waste time and effort to conduct yet another survey.
3. *An integrative approach to these infections may help reduce excessive antibiotic prescription*—this statement is not supported by the data obtained; the survey merely suggests that such an approach is popular with the paediatricians who felt like answering the questionnaire. The success of the approach in reducing excessive prescribing of antibiotics was not assessed.

Once again, we encounter a piece of survey research that is of no value to medicine, and which presents misleading conclusions. This paper is far from atypical, and it is regrettable that there are so many of these papers passing the hurdle of peer-review and subsequently polluting the academic literature.

6.3   Chiropractic: Patients' Views

Our final example is a major survey of chiropractic patients' views of chiropractic, commissioned by the General Chiropractic Council (GCC), the UK government's regulatory body for chiropractors (MacPherson et al. 2014). A total of 544 patients were recruited, and completed a questionnaire. Some of the findings are of ethical interest:

- 15% of all patients did not receive information about possible adverse effects of their treatment.
- 20% received no explanations why investigations such as X-rays were necessary and what risks they carried.
- 17% were not told how much their treatment would cost during the initial consultation.
- 38% were not informed about complaint procedures.
- 9% were not told about further treatment options for their condition.
- 18% said they were not referred to another health care professional when the condition failed to improve.
- 20% noted that the chiropractor did not liaise with the patient's GP.

Perhaps the most remarkable finding from the report is the unwillingness of chiropractors to co-operate with the GCC which, after all, is their regulating body. The researchers managed to recruit only ca. 10% of all UK chiropractors which is more than disappointing. This low response rate will inevitably impact on the validity of the results and the conclusions. It can be assumed that those practitioners who did volunteer are a self-selected sample and thus not representative of the UK chiropractic profession; they might be especially good, correct or obedient. This, in turn, also applies to the sample of patients recruited for this research. If that is so, the picture that emerged from the survey is likely to be far too positive. In any case, with a response rate of only ca. 10%, any survey is next to useless.

7 Conclusions

Since we have set out the myriad of deficiencies in CAM research, the reader might be forgiven for assuming that we believe that more RCTs ought to be conducted to examine the effectiveness (or otherwise) of various CAM modalities. Certainly, we consider that where CAM clinical research is undertaken, it ought to be carried out in a rigorous manner. However, we also consider that much research into CAM is ethically objectionable over and above questions of its quality.

Advocates of research into CAM derive support from the notion that some medically important agents or therapies may remain undiscovered, if CAM treatments are not fully investigated. Although this claim appears *prima facie* correct, it needs to be tempered in several ways. Firstly, credible scientific assessments of CAM treatments already exist. Although absolute disproof of effectiveness is rarely attainable, a reasonable estimate of plausibility and effectiveness can usually be made for most CAM modalities. Secondly, investment in research into implausible proposals implies that there is likely to be something medically worthwhile to investigate, or to be gained from such research. Thus, the act of or demand for formal investigation automatically lends a degree of undeserved respectability to an implausible claim. We do believe that this is unwarranted in many cases. In addition, results are rarely clear-cut in any one experiment or clinical trial. Anomalous or chance results in individual studies—which as described at the start of this chapter are frequent in CAM research—produce a real danger of lending credence to an implausible therapy (Ramey and Sampson 2001). Thus, we suggest that clinical trials for CAM ought to be limited to cases that do not directly violate accepted scientific laws and criteria of logic. As with mainstream medical research, a convincing database from pre-clinical research should be accumulated before patients are experimented upon.

To take another example, while it might be worthwhile trying to discover and purify the active molecular entities in certain apparently effective herbal preparations (if pre-clinical studies support this effect), it would be inappropriate to conduct large clinical trials of homeopathic remedies, considering that homeopathy is based on grossly implausible beliefs. Evaluations of such incredible systems and methods can always be performed in the (unlikely) event that new information arises from basic research.

Spurious results are frequently paraded by CAM advocates in support of implausible treatments. As discussed in the previous chapter, the more poorly conceived and executed a research project is, the more likely it is to produce false-positive results. These results then may lead to repetitive cycles of unproductive work to explain what was found—often to simply disprove the erroneous results. This is an unfortunate feature of various fields of scientific research, but it has particularly serious implications in medical research. Moreover, researchers who practice and behave as advocates of CAM may unintentionally or deliberately distort or exaggerate weak findings. Invalid CAM research claims tend not to be put to rest; instead they are repeatedly recycled.

Allocation of resources represents another important ethical issue. Rigorous trials cost between hundreds of thousands to several million pounds each, and between 5 and 20 trials may be needed to establish by systematic review the effectiveness of each product or method. Expenditure of such amounts on CAM practices such as psychic healing or homeopathy would be ethically unacceptable, given (a) the low likelihood of obtaining patient-benefiting results, (b) the effective diversion of funds from more plausible medical projects and (c) the current, largely negative evidence-base for such forms of CAM.

The National Center for Complementary and Integrative Health (NCCIH) in the USA provides a salient example of the inadvisability of CAM research. Funded by the U.S. government, the NCCIH has been directed by Congress to study CAM methods. In 25 years of existence[12] and more than $200 million in expenditures, it has added very little of value in disproving CAM methods, and has not proved the effectiveness of a single 'alternative' method. However, one of its effects clearly was to promote the CAM movement by assuming—in spite of logico-scientific arguments and evidence to the contrary—that bizarre and implausible claims are legitimate for study.

In sum, clinical research into CAM is fraught with ethical problems. While a few areas may be worthy of further exploration, in most cases the consequences of research are likely to be ethically negative, with scarce resources being wasted, false medio-scientific leads being generated, and unwarranted credence being lent to the methods. We suggest that it is ethically necessary to assume a 'default' position in which CAM clinical research is deemed inadvisable, except where on a case-by-case basis a convincing argument has been made to the effect that the above-mentioned deleterious consequences are outweighed by a sufficiently high probability of good outcomes.

On this basis, the question becomes: Which criteria should apply to evaluating proposals of clinical CAM research? The short answer is: the same criteria used in any other medical research. A prime criterion must be logico-scientific validity. If a claimed treatment is based on principles that are incompatible with well-established scientific facts, then clinical investigation of the treatment is unwarranted, because the negative features of CAM research will apply, with little realistic prospect of obtaining positive results. In response, it may be claimed that by not conducting such research, the possibility of discovering important new scientific principles would be closed off. However, this claim is based on an erroneous view of the scientific process. As we shall discuss in Chapter 6, which considers issues of truth and truthfulness (and their converse) in CAM, although scientific theories are not set in stone, and occasional 'paradigm shifts' have occurred in the history of science, significant change to a basic scientific principle requires substantive scientific evidence, garnered by extensive and painstaking research involving large numbers of scientists.

---

[12]Under a variety of different names.

Surprising results from individual studies, conflicting with core scientific precepts, can be explained in one of two ways: either (a) the relevant fundamental scientific theory is flawed and requires overhauling, or (b) there is a problem with the research itself—for example chance results generated through statistical variation, inadequate controls, post hoc selection of end points, a failure to apply appropriate controls, or even fraud. It is self-evident that explanation (b) is immensely more probable than (a). Put into slightly more philosophical terms, Occam's razor mandates that explanation (b) ought to be accepted, as it is the simplest hypothesis (i.e. the one that contains the fewest assumptions) that is able to explain the available facts as well as could any other more complex explanation.

Applying this logic to the previously mentioned example of homeopathy, it is clear that clinical research into *any* proposed homeopathic treatment is unwise. At best, such research will generate negative or conflicting results, which are of minimal interest to medical journals and other disseminators of research findings, and which in any case will be ignored or defamed by homeopathy advocates. Where positive results are occasionally obtained, the most likely explanations are either chance, or that a problem exists with the study methodology. To appeal instead to a supposed flaw in core scientific theory as an alternative explanation would be outlandish: such an attempt would necessitate the rejection of—amongst other principles—(a) the particulate theory of matter and (b) the mathematics of serial dilution. However, in the situation where a homeopathic treatment has been tested and false positive results obtained, at this point the ethically negative aspects of CAM research are bound to follow, with unwarranted credibility being afforded to homeopathy (Jacobs et al. 2003). It follows that clinical research into homeopathy and similarly implausible methods is ethically imprudent.

On the basis that logico-scientific validity ought to be the key criterion for assessing proposals for CAM research, the majority of CAM modalities would be rendered invalid for clinical research on humans. Certainly, this proscription would clearly apply to some of the most widespread CAM systems, including not only homeopathy, but also craniosacral therapy, crystal therapy, faith healing, reflexology, and iridology and many others.

This conclusion may appear radical to some, but it is not, as it is derived directly from the ethical precepts above. Disregarding this conclusion would establish double standards in medical research and entail the absurd position of endorsing research into claims that are simply incompatible with fundamental tenets of science and logic. Of course, the view is sometimes expressed that CAM represents a unique therapeutic province that should not be subject to the usual tenets of evidence and logic—and concomitant ethical analysis—that apply to conventional medicine. If that were true, it would be senseless to use the reason and logic of clinical trials to test CAM methods. The onus must lie with those making this kind of special plea to come forward with convincing arguments in support of their position. We know of no such arguments.

In cases where a proposed CAM treatment satisfies the logico-scientific criterion, at least one further criterion should apply. This is the likelihood of effectiveness criterion. In other words, a proposed treatment would have to do more than merely not clash with science and logic: it must also show credible promise of clinical effectiveness and safety equal to or above those in current use. For example, it is not unscientific to suppose that certain plants may contain pharmacologically active agents. It is, of course, true that many conventional therapeutic drugs were originally derived from plants. However, it would be inappropriate to simply accept the claims made by herbalists for the therapeutic properties of particular plants. Rather, there needs to be corroborating evidence for any such claim. For example, if sound anthropological observation indicated that an indigenous plant extract used by Brazilian rainforest dwellers seemed highly effective in treating a particular ailment, then further investigation would be justified. However, it should be emphasized that evidence of effectiveness from anthropological or similar studies is necessarily limited, and can at best only provide preliminary indicators of potential new research targets. Nevertheless, such or similar evidence is a prerequisite for sanctioning the investigation of any CAM claims that meet the logico-scientific validity criterion. In the absence of strong anthropological or equivalent evidence, there is simply no basis for researching any particular herbal or similar treatment. And such evidence, although inherently limited, remains better than—and cannot be replaced by—the weaker forms of 'evidence' frequently relied upon by CAM advocates, such as personal anecdotal accounts, or unsubstantiated claims contained in CAM publications.

Where a proposed treatment meets the dual criteria of (a) logico-scientific validity and (b) likelihood of effectiveness/safety, further research may be warranted. For example, the plant extract from the previous example might be fed to experimental animals for safety and effectiveness assessments; if shown to be safe and efficacious, this might be followed by human clinical trials. If the trials were successful, the plant extract could be chemically analyzed to determine whether its active component can be isolated such that this could be used as the basis for a new pharmaceutical drug. Of course, any treatment proved to be effective by rigorous and repeated clinical trials could no longer be classed as 'complementary/alternative', but would then swiftly become part of modern orthodox medicine (as has occurred with several herbal medicines already).

References

Adams B (2017) Universität Leipzig: Nachrichten. http://www.zv.uni-leipzig.de/service/kommuni kation/medienredaktion/nachrichten.html?ifab_modus=detail&ifab_id=1404. Accessed 2 Dec 2017

Al Bedah A, Khalil M, Elolemy A et al (2015) The use of wet cupping for persistent nonspecific low back pain: randomized controlled clinical trial. J Altern Complement Med 21(8):504–508. doi:10.1089/acm.2015.0065

Bamford JT, Ray S, Musekiwa A et al (2013) Oral evening primrose oil and borage oil for eczema. Cochrane Database Syst Rev (4):CD004416. doi:10.1002/14651858.CD004416.pub2

Beghi GM, Morselli-Labate AM (2016) Does homeopathic medicine have a preventive effect on respiratory tract infections? A real life observational study. Multidiscip Respir Med 11:12-016-0049-0. eCollection 2016. doi:10.1186/s40248-016-0049-0

Ben-Arye E, Samuels N, Schiff E et al (2015) Quality-of-life outcomes in patients with gynecologic cancer referred to integrative oncology treatment during chemotherapy. Support Care Cancer 23(12):3411–3419. doi:10.1007/s00520-015-2690-0

Beuth J, Schneider B, VAN Leendert R et al (2016) Large-scale survey of the impact of complementary medicine on side-effects of adjuvant hormone therapy in patients with breast cancer. In Vivo 30(1):73–75. doi:30/1/73 [pii]

Bronfort G, Hondras MA, Schulz CA et al (2014) Spinal manipulation and home exercise with advice for subacute and chronic back-related leg pain: a trial with adaptive allocation. Ann Intern Med 161(6):381–391. doi:10.7326/M14-0006

Bukowski EL (2015) The use of self-Reiki for stress reduction and relaxation. J Integr Med 13 (5):336–340. doi:10.1016/S2095-4964(15)60190-X

Bussing A, Ostermann T, Majorek M et al (2008) Eurythmy therapy in clinical studies: a systematic literature review. BMC Complement Altern Med 8:8-6882-8-8. doi:10.1186/1472-6882-8-8

Carroll RT (2017a) Anthroposophic medicine. The Skeptic's dictionary. http://skepdic.com/anthroposophicmedicine.html. Accessed 14 Feb 2017

Carroll RT (2017b) Cupping. The Skeptic's dictionary. www.skepdic.com/bachflower.html. Accessed 13 Feb 2017

Carroll RT (2017c) Faith healing. The Skeptic's dictionary. http://www.skepdic.com/faithhealing. html. Accessed 13 Feb 2017

Carroll RT (2017d) Naturopathy. The Skeptic's dictionary. http://skepdic.com/natpathy.html. Accessed 14 Feb 2017

Castro-Sanchez AM, Lara-Palomo IC, Mataran-Penarrocha GA et al (2016) Benefits of craniosacral therapy in patients with chronic low back pain: a randomized controlled trial. J Altern Complement Med 22(8):650–657. doi:10.1089/acm.2016.0068

Cha KY, Wirth DP (2001) Does prayer influence the success of in vitro fertilization-embryo transfer? Report of a masked, randomized trial. J Reprod Med 46(9):781–787

Chaiet SR, Marcus BC (2016) Perioperative *Arnica montana* for reduction of ecchymosis in rhinoplasty surgery. Ann Plast Surg 76(5):477–482. doi:10.1097/SAP.0000000000000312

Charkhandeh M, Talib MA, Hunt CJ (2016) The clinical effectiveness of cognitive behavior therapy and an alternative medicine approach in reducing symptoms of depression in adolescents. Psychiatry Res 239:325–330. doi:10.1016/j.psychres.2016.03.044

Cherkin DC, Sherman KJ, Balderson BH et al (2016) Effect of mindfulness-based stress reduction vs cognitive behavioral therapy or usual care on back pain and functional limitations in adults with chronic low back pain: a randomized clinical trial. JAMA 315(12):1240–1249. doi:10.1001/JAMA.2016.2323

Dai YK, Zhang YZ, Li DY et al (2017) The efficacy of Jianpi Yiqi therapy for chronic atrophic gastritis: a systematic review and meta-analysis. PLoS ONE 12(7):e0181906. doi:10.1371/journal.pone.0181906

Deng M, Wang X (2016) Acupuncture for amnestic mild cognitive impairment: a meta-analysis of randomised controlled trials. Acupunct Med 34(5):342–348. doi:10.1136/acupmed-2015-010989

Dong B, Chen Z, Yin X et al (2017) The efficacy of acupuncture for treating depression-related insomnia compared with a control group: a systematic review and meta-analysis. Biomed Res Int 2017:9614810. doi:10.1155/2017/9614810

Ernst E (2012) How to fool people with clinical trials. Edzard Ernst|MD, PhD, FMedSci, FSB, FRCP, FRCPEd. http://edzardernst.com/2012/11/how-to-fool-people-with-clinical-trials/. Accessed 27 Jan 2017

Ernst E (2015) Craniosacral therapy: wild assumptions, flimsy science and wrong conclusions. Edzard Ernst|MD, PhD, FMedSci, FSB, FRCP, FRCPEd. http://edzardernst.com/2015/11/craniosacral-therapy-wild-assumptions-flimsy-science-and-wrong-conclusions/. Accessed 29 Jan 2017

Ernst E, Schmidt K (2005) Ukrain - a new cancer cure? A systematic review of randomised clinical trials. BMC Cancer 5:69. doi:1471-2407-5-69 [pii]

Ernst E, Pittler MH, Wider B (2006) The desktop guide to complementary and alternative medicine: an evidence-based approach, 2nd edn. Mosby, St Louis, MO

Flamm B (2004) The Columbia University 'Miracle' study: Flawed and Fraud. Skeptical Inquirer 28(5):2017

Gok Metin Z, Ozdemir L (2016) The effects of aromatherapy massage and reflexology on pain and fatigue in patients with rheumatoid arthritis: a randomized controlled trial. Pain Manage Nurs 17(2):140–149. doi:10.1016/j.pmn.2016.01.004

Hejl M (2012) Zwei Festnahmen nach schwerem gewerbsmäßigen Betrug in Wien. Bundeskriminalamt. http://www.bmi.gv.at/cms/BK/presse/files/Ukrain_040912.pdf. Accessed 2 Dec 2017

Jacobs J, Jonas W, Jimenez-Perez M et al (2003) Homeopathy for childhood diarrhea: combined results and metaanalysis from three randomized, controlled clinical trials. Pediatr Infect Dis J 22(3):229–234. doi:10.1097/00006454-200303000-00005

Lambrigger-Steiner C, Simoes-Wust AP, Kuck A et al (2014) Sleep quality in pregnancy during treatment with *Bryophyllum pinnatum*: an observational study. Phytomedicine 21(5):753–757. doi:10.1016/j.phymed.2013.11.003

Lesi G, Razzini G, Musti MA et al (2016) Acupuncture as an integrative approach for the treatment of hot flashes in women with breast cancer: a prospective multicenter randomized controlled trial (AcCliMaT). JCO 34(15):1795–1802. doi:10.1200/JCO.2015.63.2893

Li X, Xiao QQ, Ze K et al (2015) External application of traditional Chinese medicine for venous ulcers: a systematic review and meta-analysis. Evid Based Complement Altern Med 2015:831474. doi:10.1155/2015/831474

Li J, Zhang S, Zhou R et al (2017) Perspectives of traditional Chinese medicine in pancreas protection for acute pancreatitis. World J Gastroenterol 23(20):3615–3623. doi:10.3748/wjg.v23.i20.3615

Lopez-Arrieta JM, Birks J (2000) Nimodipine for primary degenerative, mixed and vascular dementia. Cochrane Database Syst Rev (2)(2):CD000147. doi:CD000147 [pii]

Lotzke D, Heusser P, Bussing A (2015) A systematic literature review on the effectiveness of eurythmy therapy. J Integr Med 13(4):217–230. doi:10.1016/S2095-4964(15)60163-7

Ludtke R, Hacke D (2005) On the effectiveness of the homeopathic remedy *Arnica montana*. Wien Med Wochenschrift 155(21–22):482–490. doi:10.1007/s10354-005-0227-8

Ma L, Wang B, Long Y et al (2017) Effect of traditional Chinese medicine combined with Western therapy on primary hepatic carcinoma: a systematic review with meta-analysis. Front Med 11 (2):191–202. doi:10.1007/s11684-017-0512-0

MacPherson H, Newbronner E, Chamberlain R et al (2014) Patients' experiences and expectations of chiropractic care: a national cross-sectional survey. Chiropractic Manual Ther 23(3):2017

Nayak C, Oberai P, Varanasi R et al (2013) A prospective multi-centric open clinical trial of homeopathy in diabetic distal symmetric polyneuropathy. Homeopathy 102(2):130–138. doi:10.1016/j.homp.2013.02.004

Nordfors JM (2017) Why is not Ukrain approved yet in Austria/EU? Medic debate. http://medicdebate.org/?q=node/294. Accessed 2 Dec 2017

NCH (2017) About Us. National Center for Homeopathy. http://www.homeopathycenter.org/about-us. Accessed 30 Dec 2017

Oberg EB, Bradley R, Cooley K et al (2015) Estimated effects of whole-system naturopathic medicine in select chronic disease conditions: a systematic review. Altern Integr Med 4 (2):2017–2192

Ramey DW, Sampson W (2001) Review of the evidence for the clinical efficacy of human acupuncture. Sci Rev Altern Med 5(4):195–201

Rivas-Suarez SR, Aguila-Vazquez J, Suarez-Rodriguez B et al (2017) Exploring the effectiveness of external use of bach flower remedies on carpal tunnel syndrome: a pilot study. J Evid Based Complement Altern Med 22(1):18–24. doi:2156587215610705 [pii]

Salatino S, Gray A (2016) Integrative management of pediatric tonsillopharyngitis: an international survey. Complement Ther Clin Pract 22:29–32. doi:10.1016/j.ctcp.2015.11.003

Shi Z, Song T, Wan Y et al (2017) A systematic review and meta-analysis of traditional insect Chinese medicines combined chemotherapy for non-surgical hepatocellular carcinoma therapy. Sci Rep 7(1):4355-017-04351-y. doi:10.1038/s41598-017-04351-y

Sreenivasmurthy SG, Liu JY, Song JX et al (2017) Neurogenic traditional Chinese medicine as a promising strategy for the treatment of Alzheimer's disease. Int J Mol Sci 18(2). doi:10.3390/ijms18020272. E272 [pii]

Teitelbaum A (2017) OTC homeopathic medicines: consumer use and satisfaction. ALPHA 2016. https://apha.confex.com/apha/144am/meetingapp.cgi/Paper/354764. Accessed 02 Dec 2017

Vernon H, Borody C, Harris G et al (2015) A randomized pragmatic clinical trial of chiropractic care for headaches with and without a self-acupressure pillow. J Manipulative Physiol Ther 38 (9):637–643. doi:10.1016/j.jmpt.2015.10.002

Zeng L, Zou Y, Kong L et al (2015) Can Chinese herbal medicine adjunctive therapy improve outcomes of senile vascular dementia? Systematic review with meta-analysis of clinical trials. Phytother Res 29(12):1843–1857. doi:10.1002/ptr.5481

Chapter 4
Education

High quality education is essential for the development of competent practitioners in all medical and medically related fields. Healthcare education entails the provision of knowledge in the sciences relevant to medicine and clinical practice, coupled with appropriate practical instruction. The basic education of healthcare professionals must be designed and delivered to ensure that newly qualified practitioners are competent to both practice in their field and be able to properly comprehend and respond to future developments within and intersecting their discipline. Beyond the initial phase of inculcating basic competence, education should be a continuing process that healthcare professionals engage with throughout their professional career.

An ethical responsibility falls on all organisations involved in the education of health professionals—including universities, colleges, and professional bodies—to ensure that new healthcare recruits are sufficiently and appropriately educated. Additionally, qualified practitioners have a professional and ethical duty to engage in ongoing education. Where education is inappropriate or of low quality, serious ethical issues arise: patients treated by poorly educated healthcare practitioners are likely to be exposed to therapies, diagnostic procedures and advice that are suboptimal, ineffective or unnecessarily hazardous.

This chapter begins by looking at the ethical principles relevant to CAM education. We then set out the fundamental elements of good education in healthcare, and go on to ask whether it possible for CAM education to conform to these elements. We argue that the answer often is 'no'; certain unavoidable features of CAM education render it an almost intrinsically invalid enterprise. The remainder of the chapter considers CAM education as it currently occurs in real life, where several serious problems are evident, all of which raise substantial ethical concerns. We suggest that CAM education frequently displays similar attitudes towards its students as a religious cult displays towards its followers. We also suggest that CAM education expropriates and corrupts core concepts from mainstream medical

© Springer International Publishing AG 2018
E. Ernst and K. Smith, *More Harm than Good?*
https://doi.org/10.1007/978-3-319-69941-7_4

education, including holism and evidence-based medicine (EBM). Finally, we discuss the current tendency of some universities to offer academic degrees in CAM, and explore the inclusion of CAM education within mainstream medical curricula.

1 Education and Ethics

The ethical principles mentioned previously are applicable to medical education as they are to other aspects of medicine. These principles give rise to ethical duties which apply to both those responsible for providing education and to the students.

The principle of autonomy—which, as we previously argued, is implied by utilitarianism—gives rise to two imperatives of relevance here: respect and consent. In terms of *respect*, an ethical responsibility rests with education providers to ensure that syllabi are designed to impart reliable knowledge and effective, evidence-based skills. Deficiencies in such curricular fundamentals, whether by omission or (worse) by the promulgation of invalid knowledge, is tantamount to a failure to respect the student.

Moreover, students typically commit very substantial time and resources to their chosen programme of study (several years in the case of degree-level courses), and thus cannot easily withdraw from their commitment, if they later discover that what they are studying is of questionable validity. So, prospective students need to be provided with accurate and sufficient information upon which to base their choice of studies. If such information is inadequate or defective, it can be argued that 'informed consent' (to enrol on a programme) has not been obtained. Just as obtaining consent from patients is an ethical cornerstone in medical practice, its importance in the context of education cannot be overstated.

Regrettably, as we shall demonstrate, CAM education is frequently aberrant in terms of curricular content, and it is likely that at least some CAM students will have committed to a programme of study without fully understanding its questionable nature. In such cases, by breaching the imperatives of respect and consent, those responsible for providing defective CAM education have violated the overarching ethical principle of autonomy.

Infliction of harm is another major ethical issue in this scenario. If a CAM student has been taught a wealth of invalid knowledge, and trained in a therapeutic modality that is intrinsically unable to do better than placebos, it seems reasonable to hold that the student has been harmed. Certainly, such students have not been as well informed or well trained as they might have been, and this is an ethical issue.

In the above scenario, the nonconsequentialist ethical principle of justice will likely also be violated, in that the students concerned will have been denied the sort of valid education that they would otherwise have received had they enrolled upon a different programme of study.

It is not only students who are affected when aberrant CAM education violates the fundamental principles of ethics. The patients ultimately treated by the practitioners produced by anomalous syllabi will also be affected. Autonomy and justice will not be respected where patients 'consent' to treatment on the basis of misinformation, and patients may be harmed—directly or by failure to obtain valid treatment—by the application of ineffective or dangerous therapies.

Similarly, misinformed students who become CAM researchers are likely to cause harm by undertaking futile research and concomitantly poisoning the stock of medical knowledge. In general, the more that CAM students are taught to believe in falsehoods, the greater will be the negative impact on society.

2   Good Education

Certain features of good education are universal across all disciplines of healthcare. We suggest that the following elements should be present in all educational programs designed to create qualified health professionals:

1. Fundamental science underpinning health and disease;
2. Discipline-specific medical science;
3. Medical research skills including statistics and critical thinking;
4. Diagnostic and therapeutic skills.

This list is not all-inclusive; for example, some healthcare disciplines may require substantive education in additional areas such as social science. However, any program that omits or neglects attention to any of the above four elements will inevitably produce practitioners that are deficient in relevant knowledge or skills.

It is important to distinguish between healthcare professionals who [a] are expected to make autonomous diagnostic judgements and treatment decisions from [b] those whose jobs are restricted to carrying out a set medical procedures that do not require them to behave autonomously in a clinical sense. Examples of the latter type of roles include healthcare staff whose job it is to take X-rays, operate life-support equipment, or run laboratory tests on blood samples. Even where the procedures entailed are complex and technically demanding, roles of the latter type do not absolutely require all four of the above educational elements.[1] In this chapter, we shall restrict the terms 'healthcare professional' and 'practitioner' to roles that entail making autonomous diagnostic judgements and therapeutic decisions.

---

[1] Although such 'non-autonomous' healthcare roles do not absolutely require education in all four elements, in practice the trend in recent times has been towards providing more extensive education for these roles, with the requirement of degree-level qualification becoming commonplace. Notwithstanding the additional costs entailed, and insofar as the four key elements are covered (which is not always the case), such extended education is likely to be of value for those staff who are promoted to senior levels involving an input to clinically relevant decision-making (for example in terms of which diagnostic equipment should be purchased).

Each of the above four elements is essential for the education of all such (autonomous) healthcare professionals. Without the first element, namely sufficient in-depth knowledge of fundamental science underpinning health and disease, the healthcare practitioner cannot be expected to understand how the body functions at all the relevant levels (including systemic, haematological, tissue, organ, immunological, cellular, molecular, and biochemical levels). If the practitioner lacks such understanding, they will not be able to properly understand dysfunction and disease beyond a superficial level, nor how therapeutic interventions work.

Of course, it would be unnecessary and inappropriate to demand that all healthcare professionals receive the same breadth or depth of basic medical science education. For example, a medical student needs to have a wide scientific knowledge, since the scope of medicine includes all aspects of the body in health and disease. By contrast, many healthcare professions do not require such a thorough grounding in basic science. For example, a student physiotherapist requires sufficient basic science to understand the development and functioning of bones, joints and muscles, given their clinical focus. Nevertheless, while physiotherapists require concomitantly less knowledge than physicians of other aspects of the human body, such knowledge is still important, albeit in a relatively dilute form. The body is a complex entity with a high degree of interaction between its various systems; this dictates that specific healthcare modalities (such as physiotherapy) require a broad education in basic medical science. The study of only isolated aspects of medical science would not produce true understanding. This can be illustrated by extending on the physiotherapy example: the physiological functioning of a joint is dependent upon underlying cellular and biochemical functioning, and therefore the student physiotherapist must have an understanding of basic cell science and biochemistry. And because wider dysfunctions can cause disease in the areas of interest to the physiotherapist, the student physiotherapist must learn about the general functioning of the body. For example, because immunological disturbances can cause damage to joints, the physiotherapy student needs to learn a modicum of the science of immunology.

Knowledge of basic medical science is also required as a necessary foundation for the second element on the above four elements, namely discipline-specific scientific knowledge. In terms of this more advanced element, it is clear that healthcare practitioners who lack education in the specific science behind their discipline will not be able to understand in depth the pathology and treatment of diseases pertinent to their specialised area. Of course, the degree of specialist scientific knowledge required will depend upon the profession in question. For example, psychiatrists need to know much more about neurophysiology than obstetricians do, and vice versa for reproductive physiology.

The third essential element of healthcare education, namely medical research skills, including in statistics and critical thinking, applies to all practitioners, not merely those destined to undertake substantive research within their discipline. The practitioner lacking such skills will be unable to discern reliable knowledge from misleading and false claims. This is a serious issue: in the previous chapter, we saw that the CAM research literature is polluted with a vast number of papers that are flawed and unreliable. Practitioners who are not able to critically evaluate medical research will be unable to discern between promising and invalid findings, and therefore their practice is less likely to be evidence-based. This can only be detrimental to their clinical practice and ultimately to their patients.

Diagnostic and therapeutic skills form the fourth essential element of healthcare education. Such skills are required to ensure that the practitioner is able to: [a] accurately and reliably determine the underlying cause of a patient's symptoms; [b] choose the most effective therapy for the patient; and [c] deliver the necessary treatment in an effective and safe fashion.[2]

Because deficiencies in clinical skills can directly lead to negative or even catastrophic consequences for the patient, this fourth educational element has the most obvious ethical significance. However, the other three elements are nevertheless important from an ethical perspective, because ultimately it is these educational foundations that will underpin the professional's clinical judgement, performance and behaviour.

3   CAM Education: Can It Ever Be Valid?

How does the education of CAM practitioners measure up to the above four essential elements? The answer is: in general, very badly. Later in this chapter we shall give some specific examples of the various types of inadequate and flawed education that predominate in the world of CAM. Before doing so, however, we shall consider how CAM programs frequently fail to live up to the essential requirements of medical education. As will become evident, the nature of CAM per se often means that meeting the basic requirements of sound medical education is inherently unattainable.

The first point to make is that, in contrast to most mainstream healthcare modalities, CAM education is frequently unregulated by government. Consequently, CAM programs vary widely—even when a single modality is considered. For example, in the UK there are several programs through which a student can become a homeopath, with these courses varying greatly in many respects, most obviously in terms of their duration (disturbingly, some UK homeopaths have no healthcare-related education at all). Similarly, the German Heilpraktiker (healing practitioner), a non-medically qualified CAM practitioner, might have several years of formal training or no training at all. Clearly, programs that have a very short duration cannot possibly cover the essential elements of medical education. Some CAM programs do run for a sufficient length of time— several years in the case of some CAM degrees—to potentially allow coverage of

---

[2]In practice, there will often be a division of labour, especially in modern mainstream medicine; so the professional who makes the diagnosis (of kidney stones, for instance) may not be the same person who performs the treatment (i.e. the surgeon who removes the kidney stones).

the four essential educational elements. However, these programmes are also prone to being fundamentally flawed in a number of other respects.

3.1 Deficits in Fundamental Science Education

In terms of education in the basic science underpinning health and disease, CAM programmes tend to take one of two approaches, both of which are problematic: [a] only a minimal degree of basic science is included, which is insufficient as a means to properly understand the human body in health and disease; or [b] a decent quantity of science is covered, but with this being largely or completely detached from the CAM modality.

Little needs to be said about the former approach; clearly, programs with insufficient basic medical science content simply cannot provide a robust foundation for the education of healthcare professionals. This is obviously problematic from an ethical perspective. By contrast, the latter approach, which typically involves students in their early years being co-taught alongside students in mainstream disciplines, at least provides the possibility of CAM students gaining sufficient basic science knowledge. However, the risk with this model is that the imparted scientific knowledge is likely to be treated as abstract from, and therefore largely irrelevant to, clinical aspects of the CAM modality being studied. This intellectual detachment may be mitigated by attempts within programs to actively cross-refer basic science teaching with healthcare and discipline-specific components. Nevertheless, to the extent that a lack of genuine integration occurs, such co-taught models of education are ethically questionable.

3.2 Deficits in Discipline-Specific Medical Science Education

More substantive problems arise with the parts of the CAM syllabus where discipline-specific medical science is covered. Here there is often a fundamental problem: the claimed mechanisms by which many CAM modalities function are often completely antithetical to science itself! We have previously referred to the supposed modus operandi of two of the most prominent forms of CAM, acupuncture and homeopathy. These CAM modalities serve as apposite examples of the intrinsic problems with attempts to 'educate' students in the specific mechanisms of CAM.

Consider traditional acupuncture: educating students to 'understand' how acupuncture affects the body entails teaching completely implausible physiological features and mechanisms, including Qi (life energy) flowing through meridians (body channels). No such 'energy' has ever been detected—despite decades of advanced scientific research into physiological communication and signalling systems; nor have any acupuncture 'channels' been revealed, despite decades of painstaking research into the anatomy and physiology of human tissues and organs. Our scientific understanding of the human body is, of course, not yet complete, but no gaps in our understanding exist that require explanations based on Qi or meridians.

Similarly, homeopathy supposedly operates on principles that fly in the face of copper-bottomed scientific knowledge concerning not only human physiology but also fundamental tenets of chemistry and physics. In order to 'understand' the mechanisms of homeopathy, a student would need to believe the following anti-scientific notions: that a substance which causes particular symptoms can be used to treat an illness with similar symptoms; that medicines so dilute as to contain no active molecules can nevertheless exert therapeutic effects; that vigorous shaking activates a vital force in homeopathic medicines; and that the solvent used for the dilution of these medicines retains a 'molecular memory' of the original substance.

Some students may come to see through and reject such pseudoscientific assumptions. These individuals, from whose eyes the scales have fallen, will find themselves in a dilemma: they may already have invested a great deal of time and money in the CAM program, and by quitting they would see their investment written-off, and the end to the possibility of following a career as a CAM practitioner (which presumably was their purpose in enrolling on the program). Such students must either leave the program and start over, or continue with their studies in complete intellectual 'bad faith'. Thus, the imposition of this dilemma, which is entirely the fault of CAM programmes and those who run them, directly harms students, either by wasting their time and money (and quite possibly damaging their future career prospects) or by incentivising them to learn and reiterate falsehoods in order to graduate.

3.3   Deficits in Research Skills Education

Just as CAM education programs would be undermined by the inclusion of too much substantive science content, a genuine focus on teaching research skills would likely lead all but the most blinkered students to realise that much of CAM rests upon extremely shaky foundations. Accordingly, such skills, including statistics and critical thinking, are either excluded from CAM programs or, if they are present, are taught in aberrant forms - frequently by CAM enthusiasts who reject core principles of clinical research.

For example, CAM students are frequently taught that RCTs—which are generally viewed as a 'gold standard' by mainstream researchers—represent merely one form of evidence which can where convenient be replaced by other, more CAM-friendly forms of 'evidence', such as uncontrolled observational studies or even anecdotal accounts from individual patients.

As well as blinding students to the truth, and thus impacting negatively on their ability to respond appropriately to future claims and critiques pertinent to their field, the failure to inculcate the necessary intellectual skills is liable to produce CAM researchers who will design, conduct and condone forms of research that are prone to generating misleading conclusions. One only need refer back to the examples of CAM papers in the previous chapter, and observe the low quality, weak and flawed nature of these studies, in order to see the reality of how CAM proponents often are educated to conduct and interpret their research.

3.4   Deficits in Diagnostic and Therapeutic Skills Education

The final element of good healthcare education, diagnostic and therapeutic skills, can only be meaningfully developed if the form of CAM being studied is inherently valid. In general, the ways in which CAM considers disease and dysfunction to arise are very different from the mechanisms understood by medical science. So, instead of specific factors (such as mutations or infections) as the cause for particular disorders, CAM modalities typically posit a universal phenomenon as the underlying cause of a broad range of disorders, or even of all illnesses.

Examples include 'imbalances' in Qi as found in acupuncture (and many other CAM modalities), the 'vertebral subluxations' at the heart of chiropractic, and the 'fluid imbalances' of craniosacral therapy. This 'one cause' ideology is deeply implausible and conflicts head-on with established medical knowledge. Yet, it is a hallmark of almost all forms of CAM. The notion of a single cause for disease resides in the realm of fantasy; and where patients' diagnoses are based upon such fiction, the outlook for treating medical conditions is not favourable.

A common feature of CAM education is the argument that CAM practitioners 'treat the root causes of disease'. This dictum is closely related to the 'single cause' notion of disease causality. If traditional acupuncturists, for instance, are convinced that any disease is the expression of an imbalance of life-forces, and that needling acupuncture points will re-balance these forces thus restoring health, they must automatically assume that they are treating the *root cause* of any condition. If chiropractors are educated to believe that all diseases are due to 'subluxations' of the spine, it must seem logical to them that spinal 'adjustment' is synonymous with treating the *root cause* of whatever complaint the patient is suffering from.

If these practitioners were correct in believing that their interventions are causal treatments, i.e. therapies directed against the cause of a disease and not merely its symptoms, then the therapy in question ought to completely heal the problem at hand. This obviously begs a crucial question: *are there any diseases which are reproducibly cured by a CAM therapy?* The simple answer is: no. A few CAM therapies have shown some effectiveness against specific conditions: examples include hypnosis for labour pain, melatonin for insomnia, and relaxation for anxiety —although the supporting evidence is still weak, and in most cases conventional treatments are more effective (Ernst 2008b). However, even if we accept that these CAM treatments really work, their effectiveness is not causal but symptomatic by nature. Therefore, the belief that CAM can provide causal treatments is unsupported by evidence, as well as lacking plausibility; yet this flawed concept is routinely taught to CAM students. This is another instance of ideological indoctrination masquerading as education in CAM.

Even where mainstream medicine has first established a reliable diagnosis for a patient, treating the patient with CAM is likely to be ineffective. Just as CAM therapists are wedded to their erroneous beliefs as to disease causality, they often are committed to treatments that are implausible or unproven. So, no matter how diligent, thorough and well-educated (in CAM) is the practitioner, application of CAM to diagnosis and therapy is almost certain to be of no specific real and lasting benefit to the patient. It may even harm the patient's health, either directly or through failure to seek appropriate (mainstream) treatment. The best that may be hoped for is that sufficient guidance and knowledge has been included in the CAM practitioner's education to allow the recognition of possible serious illness so that the patient can be referred for *bona fide* investigations and treatments. Yet where mainstream medical education is included in a CAM program, the whole point of the 'CAM' component is called into question, as the inclusion of non-CAM educational content implies that the CAM modality is (at best) only useful for relatively trivial disorders. This is an admission that CAM proponents are generally unwilling to make, and it is therefore unsurprising that the educational programs which such advocates design and deliver usually do *not* include substantive teaching of mainstream diagnostic and therapeutic principles. More often than not, such programs teach about conventional healthcare in a derogatory fashion defaming modern drugs for having serious side-effects, for instance.

4 Cults and CAM Education

'Educating' students to believe in the sorts of pseudoscience described above arguably amounts to indoctrination. By attempting to make students believe in fantastical notions, this kind of instruction cannot help students to develop a deeper scientific understanding: rather it is likely to shut down their desire to seek the truth. Attempting to make students believe in palpable nonsense, and follow a career based upon it, is akin to the brainwashing of recruits that occurs within religious cults. All this is deeply concerning in and of itself from an ethical perspective, because such indoctrination violates the autonomy of the student, and can only be harmful, foremost to patients.

Worryingly, CAM has some of the qualities of a cult. One characteristic of a cult is the unquestioning commitment of its members to the bizarre ideas of one single person. Homeopaths, for instance, very rarely question the implausible doctrines of its 19th century founder, Samuel Hahnemann. Similarly, few chiropractors doubt the assumptions of their founding father, D.D. Palmer. Such cult-leaders are idealised and cannot be proven wrong by logical arguments nor scientific facts. Proper education, however, must be open to new ideas and thrives on questioning established principles.

Cult members tend to be on a mission and often operate on a political level to support and popularize their cult. They cherry pick evidence, argue emotionally, adopt conspiracy theories, ignore arguments which contradict their belief system, and tend to assume that there is little worthy of their consideration outside the cult. Therapies, concepts and facts which are not cult-approved are systematically ignored or defamed. Education of cult-members thus takes the form of brain-washing, where concepts and ideas can never be scrutinised but must be accepted unconditionally. Opposition is not tolerated and critics are ousted; to belong to the cult, one must believe in its tenets, and those who fail to fulfil this criterion will be excluded from the cult.

Cult members blindly adhere to what they have been told. The effects of such pseudo-education can be dramatic: the powers of discrimination of the cult member are reduced, critical thinking discouraged, and no amount of evidence can dissuade cult members from doubting what they have learnt.

We are not suggesting that all CAM practitioners are members of a cult. However, the evidence we present below suggests that the education of CAM practitioners is frequently woefully deficient and has features in common with a cultish brainwash.

5 Expropriation of Concepts from Mainstream Medicine

The education of CAM practitioners tends to highjack concepts that are essential elements of conventional medicine, pretending that they are unique to CAM. Examples include holism, evidence-based medicine (EBM), prevention, and compassion. This expropriation of mainstream concepts is of particular relevance to CAM education: these concepts are core philosophical features of healthcare, and teaching students warped versions of them is likely to produce practitioners who are seriously deluded in their fundamental approaches to health and disease. We consider two of these CAM hijackings below, namely holism and EBM.

5.1 Holism

The concept of 'holism' refers to viewing the whole person in his or her emotional, familial and societal context, underpinned by sufficient consultation time. It is widely accepted that a holistic approach is generally beneficial for patient care. Holism can be, and often is, built into conventional healthcare. But CAM practitioners are frequently taught that they are the only clinicians who truly practice holistically. This notion implies that conventional medicine is not holistic.

Providers of CAM education frequently make this clear in their webpages or promotional material. For example, under the heading 'Why Study With Us', the UK's *School of Homeopathy* makes the following claim: "*Although current medical practice recognises that there are connections between mental, emotional and physical conditions, the dominant approach is to break the whole down into parts with separate diagnostic labels*" (School of Homeopathy 2017).

Similarly, the *British Acupuncture Accreditation Board* website presents holism as follows: "*In medical contexts, holism entails the integration and interpenetration of the life of the spirit with that of the mind and body, requiring always that the whole person be treated in sickness and in health. There is nothing irrational or antiempirical about this holism, even though it takes practitioners into areas of reality and knowledge uncharted by mainstream medical science*" (Hougham and Parrott 2017).

These examples illustrate the general stance to holism taught to CAM practitioners, which could be paraphrased as follows: while mainstream medicine may contain some holistic elements, 'real' or 'deep' holism is unique to CAM. This assertion is made by many CAM educational organisations, encompassing a wide range of modalities.

By contrast, we contend that the kind of holism to which mainstream medicine aspires—namely not merely or exclusively seeing health and disease as abstract entities separate from the whole individual—is the correct one. Nevertheless, contra the above *School of Homeopathy* quote, sometimes it *is* entirely rational, and medically effective, to "*break the whole down into parts with separate diagnostic labels*"—an approach that CAM proponents like to call 'reductionism', employed as a term of deprecation. For example, prevention, diagnosis and treatment of hepatitis B requires specialist focus on immunology, virology and hepatology. This 'reductionist' approach does not mean that holistic aspects of hepatitis B do not exist; they do, and might (depending on the individual patient) include occupational and lifestyle considerations in respect of risk factors for transmission, together with ways of coping psychologically with the disease.

Indoctrinating CAM students with the notion that conventional medicine is not holistic is quite simply misleading. This distortion is compounded when CAM education promulgates the false idea that it is wrong to 'break down' a disorder into 'separate parts'. Worse still, these defects in CAM educational philosophy are frequently compounded by the teaching of magical or mythical elements, such as (from the above quotes) "*the life of the spirit*" and "*areas of reality and knowledge uncharted by mainstream medical science.*"

This philosophy of 'deep holism' is essentially an ideological position: it is not based on rational or evidential considerations. It is important to note that this kind of holism, in which everything about the patient must be considered as a unified and indivisible whole, intermeshes with and supports the above-mentioned pseudoscientific CAM notion that disease per se has a unitary source (e.g. Qi imbalance), supposedly treatable with a single therapeutic approach.

The sort of unbridled holism often promoted in CAM can actually be problematic. While extolling its virtues, CAM proponents inevitably fail to acknowledge that holism is not without its drawbacks: overly holistic approaches risk medicalising problems that are better addressed outside the consulting room, with an associated risk of rendering patients excessively dependent upon healthcare practitioners. Doubtless some CAM practitioners might view this as a benefit, either due to a belief that maintenance of good health requires regular CAM attention (e.g. through adjustments to Qi or differently named life-forces), or simply as a means of retaining custom. In either case, the creation of dependency beyond the treatment of specific disease is ethically fraught, because dependence implies an effective lack of informed consent, and may also amount to a form of harm to the patient.

There is an irony in misleading CAM students about holism: the reality is that many forms of CAM are themselves not holistic at all. For example, there is no holism involved in the (now ubiquitous) provision of herbal and homeopathic preparations for sale online and in high-street pharmacies. And what could be less holistic than studying a patient's iris, as iridologists do, to draw far-reaching conclusion about his health from it? One wonders how the average CAM practitioner, whose training has emphasised the importance of holism, views such 'reductionist' CAM offerings.

Holism is undeniably a central and essential element of conventional health care. Yet educational institutions of CAM misinform future CAM practitioners that they have a monopoly on it. They use it to create a straw man misleading the public, and pervert it into a tool for attracting and financially exploiting the often all too gullible public.

5.2 Evidence-Based Medicine

Proponents of CAM are frequently educated to claim or imply that they practice evidence-based medicine (EBM). Their argument usually holds that EBM represents much more than just data from clinical trials, and that they do abide by the rules of EBM when treating their patients. The former claim is correct but the latter usually is not. To explain why, we ought to first define our terminology. According to David Sackett, who was part of the McMaster group that coined the term, EBM is "the conscientious, explicit and judicious use of current best evidence in making decisions about the care of individual patients. The practice of evidence-based medicine means integrating individual clinical experience with the best available external clinical evidence from systematic research" (White 2017). As proposed by Sackett, the practice of EBM rests on the following three pillars:

- **External Evidence**—clinically relevant and reliable research mostly from clinical investigations into the efficacy and safety of therapeutic interventions— in other words clinical trials and systematic reviews.
- **Clinical Expertise**—the ability to use clinical skills to identify each patient's unique health state, diagnosis and risks as well as his/her chances to benefit from the available therapeutic options.
- **Patient Values**—the individual preferences, concerns and expectations of the patient which are important in order to meet the patient's needs.

Consider a homeopath treating a patient with migraine, a chiropractor manipulating a child with asthma, or an acupuncturist needling a consumer for smoking cessation. The best available external evidence shows that none of these therapies is effective for these indications (Ernst et al. 2008). So, why do these CAM practitioners often claim to be practice EBM? The short answer is that their education has led them to discount or trivialise the external evidence.

Using the first example of the homeopath, the scenario usually is as follows: a homeopath has been educated to believe in the effectiveness of homeopathy and has the clinical expertise in it (he probably has clinical expertise in nothing else but homeopathy). His patient's preference is very clearly with homeopathy (otherwise, she would not have consulted him). It follows that the homeopath's practice is based on the latter two pillars of EBM—clinical expertise and patient values. As to the first pillar—external evidence—he is adamant that clinical trials cannot do justice to something as holistic, subtle, individualized etc. Therefore, he refuses to recognize the trial data as conclusive and rather trusts his experience or uncontrolled observational studies—both of which might be substantial.

Yet the practice of EBM must rest on all three pillars: each one is essential. We cannot just pick the ones we happen to like and drop the ones which we find awkward, we need them all. To pretend that external evidence can be substituted by something else is erroneous, and introduces double standards which are not acceptable not because this would be against some bloodless principles of nit-picking academics, but because it would not be in the best interest of the patient. After all, the primary concern of EBM must be the patient. As long as CAM practitioners are being educated in an anti-science spirit, we cannot be surprised by their often complete and dangerous misconceptions of the truth.

Steve Scrutton, a UK homeopath and a Director of the '*Alliance of Registered Homeopaths*' (ARH) which represents nearly 700 homeopaths in the UK, might serve as a good example (ARH 2017). On his website, he promotes homeopathy as a treatment and prevention for measles: "*Many homeopaths feel that it is better for children, who are otherwise healthy, to contract measles naturally. Homeopathy is less concerned with doing this as it has remedies to treat measles, especially if it persists, or become severe. Other homeopaths will use the measles nosode, Morbillinum, for prevention. Homeopaths have been treating measles for over 200 years with success.*" At the end of this webpage, Scrutton makes the following statement: "*I am not aware of any scientific studies on the homeopathic treatment of measles*" (Scrutton 2017).[3]

Why would anyone make such dangerous statements? The answer is that CAM practitioners are trained to uncritically over-estimate their therapies. In 2008, Scrutton wrote: *What 'scientific' medicine does not like about homeopathy is*

---

[3]Intriguingly, the ARH has a code of ethics which its members evidently disregard. It states that they must "not claim or imply, orally or in writing, to be able to cure any named disease and that they should be aware of the extent and limits of their clinical skills" (ARH 2017).

not the lack of an evidence base—it is the ability to help people get well—and perhaps even more important, we can do it safely.

(Sadly, some of the links to Scutton's public views went dead following the exposure of his statements in pro-science blogs.)

It is clear that CAM proponents see EBM as an important part of education. Students of CAM themselves seem in favour of EBM. This is illustrated by a study in which chiropractors collected data from chiropractic students enrolled in colleges throughout North America. The stated purpose of this study was to *investigate North American chiropractic students' opinions concerning professional identity, role and future* (Gliedt et al. 2015).

In this study, 7455 chiropractic students from 12 chiropractic colleges were invited to complete an electronic questionnaire; a total of 1243 questionnaires were returned. Most respondents (87.0%) agreed that it is important for chiropractors to be educated in evidence-based practice. However, a majority (61.4%) of the respondents agreed that the emphasis of chiropractic intervention is to eliminate vertebral subluxations/vertebral subluxation complexes—a notion which is utterly unsupported by good medical evidence.

These survey results illustrate a central conflict in CAM education: a desire for education in EBM versus the teaching of implausible therapies. There is no good evidence to support the notion that subluxation is a valid medical concept, nor that chiropractic can effectively treat disorders supposedly caused by subluxation. But evidence does exist to show that chiropractic manipulations can lead to serious injury, including quadriplegia (Ernst 2008a). So, genuinely teaching chiropractic students the principles and practice of EBM should lead them to reject subluxation— which would fundamentally undermine a central principle of traditional chiropractic medicine.

The way around such conflicts between evidence and ideology is, of course, to teach a corrupted version of EBM, in which one of the three pillars mentioned above—namely external evidence—is disregarded.

Perhaps the starkest manifestation of a failure to deal properly with medical evidence is the anti-vaccination stance adopted by many CAM commentators and practitioners. Vaccination has been one the greatest triumphs of modern medicine, and has saved countless lives that otherwise would have been lost to, or blighted by, infectious disease. Yet the anti-vaccination pundits within CAM loudly decry vaccination. Their grounds for doing so are based on groundless or exaggerated fears of health risks from vaccination; yet despite the entirely pseudoscientific nature of their claims, these vocal ideologues have succeeded in doing untold damage by dissuading parents to have their children immunized.

This anti-vaccination attitude prevails amongst CAM practitioners, including chiropractors, homeopaths, naturopaths and others. The origins of this regrettable and unethical feature of CAM stem from problems with the education of practitioners. One example comes from a survey of 560 naturopathy students at the National College of Natural Medicine in Portland, USA, about their knowledge of and attitudes towards vaccination, with some startling—but by no means atypical— findings (Ali et al. 2014).

All of the students surveyed had learned about vaccination, both via formal teaching and independent study. The information sources used for this learning had varying levels of credibility. Although 82% supported the general concept of vaccinations for prevention of infectious diseases, only 26% of the responding students planned on regularly prescribing or recommending vaccinations for their patients, and the vast majority (96%) of those who might recommend vaccinations reported that they would only recommend a schedule that differed from the standard schedule used throughout the US.

The pseudoscientific core of these students' education is reflected by various pseudoscientific beliefs ubiquitous in CAM. A common view expressed by the respondents was that 'too much' vaccination is a bad thing—a notion that is not supported by scientific theory or evidence. Specifically, most students were concerned about: vaccines being given 'too early' (73%); 'too many' vaccines being administered simultaneously (70%); and 'too many' vaccines being given overall (59%). Similarly, most of the respondents were concerned about preservatives and adjuvants in vaccines (72%)—despite the fact that such additives are present at too low a quantity to plausibly damage health, and no good evidence exists to indicate an actual health risk. About 40% believed that a healthy diet and lifestyle was more important for prevention of infectious diseases than vaccines—a simply fantastical and most dangerous notion. Overall, 90% of the responding naturopathy students admitted that they were more critical of vaccines than mainstream paediatricians, medical doctors, and medical students.

These findings indicate that naturopaths (similar findings exist for homeopaths and chiropractors) are educated to become anti-vaccinationists who believe that their CAM treatments offer better protection than vaccines. They are thus depriving many of their patients of arguably the most successful means of disease prevention that exists today. To put it bluntly: student naturopaths are brainwashed into becoming a danger to public health. Yet this is just one example; it is clear that, in general, CAM practitioners are educated to be sceptical about vaccination.

6 University Education and CAM

In contrast to the education of physicians and other conventional healthcare professions, CAM education is largely unregulated. It is therefore unsurprising to find —even for just a single modality—a vast array of CAM courses delivered by a wide range of (mostly private) providers. This laissez-faire market may hold appeal to some individuals aspiring to become CAM practitioners, in that it is possible to become 'qualified' in a CAM by selecting a short and relatively undemanding course.

However, CAM programs are also delivered by bona fide universities. And many of these courses are substantive, leading to the award of a degree. This suits CAM proponents: the more that universities are seen to teach unconventional therapies, the greater will be the perceived respectability of such therapies, given the high levels of respect accorded by the public to university education.

While some of these courses may claim to offer a 'critical' approach, the reality is that many of these programs simply advocate unconventional therapies in a non-analytical or even promotional fashion. As discussed above, it is difficult to conceive of substantive educational provision in CAM that could be based on anything other than training students to appreciate and apply a range of therapies. This inherent tendency towards the indoctrination of students in favour of anomalous therapy methods is ethically objectionable—all the more so, we suggest, when the educational body is a university, with its high social status.

CAM degrees are taught at universities across the world. A worldwide survey is beyond the scope of this book, and instead we shall focus on the situation in the UK. To a large extent, the situation in this country mirrors that in other developed nations; much the same problems have been reported in Australia (MacLennan and Morrison 2012), and informal searching of the Internet yields a huge number of hits for adverts for CAM degrees from universities in the US, Canada, Europe and beyond.

From 1992, the UK began allowing former colleges of higher education to apply for university status. This resulted in the number of universities and university-equivalent providers increasing approximately 3-fold. The new institutions were able to act autonomously, for example to launch entirely new degree programmes without the oversight of the national degree-awarding authority. This expansion and increased institutional freedom led to a substantial increase in the array of degree programmes being taught, and part of this increase included the establishment of degree programmes in CAM.

In 2007, an important article in the science journal Nature exposed this parlous situation. It revealed that the UK's central Universities and Colleges Admissions Service (UCAS) advertised 61 courses for complementary medicine, of which 45 were Bachelor of Science (BSc) honours degrees, five of which were homeopathy degrees (Colquhoun 2007). Other CAM courses included acupuncture, aromatherapy, naturopathy, nutritional therapy, osteopathy, reflexology, and traditional Chinese medicine. Despite being taught under the BSc label, none of these areas could be considered as 'science'.

Something of a furore broke out when the Nature article was published, leading to a great deal of criticism being directed towards those universities that were providing CAM degrees. This negative attention led to the closure of many of these courses. For example, by 2009 all five homeopathy degrees had ceased (Colquhoun 2017).

However, UK universities continue to teach CAM degrees—including BSc degrees. At the time of writing, UCAS was advertising 24 CAM undergraduate degrees (UCAS 2017). (This figure does not include degree variants delivered by the same institution, such as sandwich year or part-time variants, so the absolute number of CAM degrees is in fact higher than 24.) Eleven of these degrees are in chiropractic or osteopathy; they are not BSc degrees, which is one small grace; nevertheless, the two CAM modalities involved are certainly not based on sound medical evidence.

The remaining 13 CAM programmes advertised by UCAS are all BSc degrees. The named degree awards include acupuncture, Chinese medicine, complementary healthcare, herbal medicine, osteopathic medicine, and professional practice in complementary therapies—none of which could remotely be described as 'science'.

Additionally, some institutions without degree-awarding powers of their own run CAM degrees that are validated by other institutions. In some cases, these degrees are listed by UCAS, but not always. For example, the 'Centre for Homeopathic Education' offers a BSc (Hons) homeopathy degree which is validated by Middlesex University (CHE-LON 2017)—sadly marking the return of the hitherto extinct BSc homeopathy degree. The extent to which such externally validated CAM degrees exist outside the UCAS listings is unknown.

It is interesting to note that, of the UK universities providing (or accrediting) CAM degrees, all but one is a 'new' university (i.e. created by conversion from a college following the abovementioned 1992 expansion).[4] Whatever opinion one may have on the general merits and demerits of massively expanding access to higher education (as has occurred in the UK and many other countries since ca. 1990), it is in our view highly regrettable that this expansion has, in some cases, led to the creation of degrees in pseudoscience and nonsense. Happily, most new UK universities do not run CAM degree programmes; but those that do ought to ask themselves the question: is it ethical to recruit students—most of whom will be young and academically naive—to programmes that condemn the learner to 3 or 4 years of brainwashing?

Degrees in CAM are only the tip of the iceberg: health-related degrees that are not primarily or expressly focused on CAM nevertheless frequently contain substantial CAM content. Various disciplines can in this way can serve as Trojan horses for CAM. Examples include sports therapy, wellness management, midwifery, and nursing; the extent of CAM coverage within these and other similar programmes varies greatly—from zero content to several entire CAM modules—depending on the institution. In this way, the exposure to CAM received by an undergraduate healthcare student is largely matter of luck, except where the prospective student exceptionally takes extra time to inquire into the finer details of curricular content when deciding which university to apply to.

In conclusion, the UK serves as an example of the persistent tendency of some (mostly lower ranking) universities and institutions to run CAM degree programmes. This tendency can be limited by the voices of proponents of science-based medicine, as evidenced by the contraction of CAM degree provision following the abovementioned Nature revelations. But like a hydra, CAM education is able to regenerate itself following pruning. CAM ideologues press for such provision because university degrees in CAM serve as an unspoken badge of approval of what is in reality pseudoscience. And, of course, recruiting students is good business for the degree providers. But from an ethical perspective, the provision of CAM degrees is simply unacceptable.

---

[4]The one exception being Swansea University.

7   Mainstream Medical Degrees and CAM

Undergraduate medical education—namely the means by which future physicians are produced—has also not been immune to the current interest in CAM. It seems reasonable to educate tomorrow's physicians about the realities of CAM because, as practitioners, they are likely to find that a sizable proportion of their patients use or ask about CAM. A similar need arises in other healthcare professions such as nursing or midwifery; however, here we shall focus specifically on the education of medical doctors.

A straightforward awareness of CAM, taught in a sceptical and analytical manner, is desirable in medical education. However, the risk is that CAM education goes further to include, or become, uncritical pro-CAM promotion. Regrettably, this does occur in many medical curricula.

In the US, it has long been the case that advocacy and non-critical assessment of CAM are the approaches taken by most medical schools: a survey of CAM curricula in U.S. medical schools in 1995–1997 showed that of 56 course offerings related to CAM, most lacked the scepticism and critical thinking appropriate for unproven therapies; only 4 were oriented to criticism (Sampson 2001). Subsequent reports in 2002 and 2009 supported this conclusion (Marcus and McCullough 2009; Brokaw et al. 2002). The most recent report available at the time of writing indicated that half (50.8%) of all US medical schools include CAM courses (Cowen and Cyr 2015).

Wallace Sampson, the author of the above paper reporting the 1995–1997 survey, an educator at Stanford University School of Medicine, called upon all medical schools to adopt the critical approach to CAM employed at Stanford, and specifically "include in their curricula methods to analyze and assess critically the content validity of CAM claims". Sadly, despite his paper being widely discussed and cited many times, there is no reason to believe that CAM in medical schools is now taught in a more critical fashion than it was before.

In the UK, the General Medical Council (GMC) sets the standards for undergraduate medical syllabi in the UK, expressed in its document Tomorrow's Doctors —outcomes and standards for undergraduate medical education. This requires, as an educational outcome, that graduate doctors must: "Demonstrate awareness that many patients use complementary and alternative therapies, and awareness of the existence and range of these therapies, why patients use them, and how this might affect other types of treatment that patients are receiving".

This outcome seems reasonable, since it only requires the inculcation of a simple 'awareness' of CAM. However, it turns out that, in practice, this outcome has been implemented in ways that vary widely: a survey of all medical schools in the UK, to which two-fifths responded, found that the extent to which individual syllabi incorporate CAM ranged hugely, from merely a single CAM lecture in the entire degree course, to CAM being considered throughout the course, entailing many hours of study and a wide range of teaching and learning approaches (Smith 2011).

The perceived preferences of students appears to be an important influence on the extent to which CAM gets incorporated into curricula. The above study found that the views of students seemed to be pulling in opposite directions: some students were "strongly in favour" of increased coverage, while others were "adamantly opposed". Amongst faculty members, a similar degree of polarisation on the issue was reported.

Backing for CAM education amongst students and staff could simply be reflective of a wish to understand what patients may be exposed to or may ask about in the way of CAM. Alternatively, such enthusiasm could indicate an ideological favouring of CAM and/or an erroneous belief in the effectiveness of CAM therapies. It is evident from the results of the above survey that both of these contrasting motivations were present within medical schools.

Reassuringly, the above study also found that CAM education in the majority of responding medical schools appeared to comprise critically describing CAM modalities, as opposed to advocating CAM in a non-analytical fashion. Nevertheless, two potentially problematic approaches to CAM were evident in some medical schools: [a] the teaching of CAM content by CAM practitioners or CAM-specific academics; and [b] the use of student-centred assignments as a form of CAM education.

In terms of [a], it is difficult to conceive of education from CAM specialists as being based on anything other than attempting to teach students to a pro-CAM agenda. This is a form of indoctrination, and is thus of significant ethical concern. In terms of [b], it is worrying that inexperienced medical students may be uncritically assimilating information from the many pro-CAM journals and websites that exist, in the preparation of their coursework.

Of course, it is crucial that physicians are properly educated regarding the realities of CAM, and the greater the extent of such learning the better from the patients' perspective. But at least in the US and the UK, the reality is that medical students are frequently not receiving sufficiently critical and sceptical CAM education. Considering the power and responsibility that the role of physician carries, medical schools have a strong ethical duty to design their syllabi such that academic standards are observed, the principles of evidence-based medicine are not violated, and pro-CAM ideology masquerading as education is not promulgated to medical students.

8 Conclusions

CAM education contains inescapable features that render it as a fundamentally unsound and often unethical activity. In practice, CAM education often resembles a religious cult. Core mainstream medical concepts, including holism and EBM, are expropriated and distorted by CAM educationalists. Some universities provide degree level CAM education, and mainstream medical curricula are often corrupted with pro-CAM content.

CAM education is thus often in danger of promulgating mistruths, pseudo-science and warped thinking amongst its students. This inevitably violates the core ethical principles of utility, autonomy and justice. Most importantly, it inflicts intellectual harm on students and ultimately leads to medical harm amongst patients treated by those 'educated' as practitioners of quackery. These harms amount to a grave moral indictment of CAM education at all levels.

References

Ali A, Calabrese C, Lee R et al (2014) Vaccination attitudes and education in naturopathic medicine students. J Altern Complement Med 20(5):A115–A116

ARH (2017) What is the ARH. Alliance of registered homeopaths. http://www.a-r-h.org/. Accessed 10 Mar 2017

Brokaw JJ, Tunnicliff G, Raess BU et al (2002) The teaching of complementary and alternative medicine in US medical schools: A survey of course directors. Acad Med 77(9):876–881. doi:10.1097/00001888-200209000-00013

CHE-LON (2017) BSc (Hons) homeopathy part time. The Centre for Homeopathic Education. http://uk.chehomeopathy.com/what-we-offer/practitioner-training/uk-homeopathic-courses/. Accessed 16 Mar 2017

Colquhoun D (2007) Science degrees without the science. Nature 446(7134):373–374. doi:10.1038/446373a

Colquhoun D (2017) The last BSc (Hons) homeopathy closes! But look at what they still teach at Westminster University. DC's improbable science. http://www.dcscience.net/2009/03/30/the-last-bsc-hons-homeopathy-closes-but-look-at-what-they-still-teach-at-westminster-university/. Accessed 16 Mar 2017

Cowen VS, Cyr V (2015) Complementary and alternative medicine in US medical schools. Adv Med Educ Pract 6:113–117. doi:10.2147/AMEP.S69761

Ernst E (2008a) Chiropractic: a critical evaluation. J Pain Symptom Manage 35(5):544–562. doi:10.1016/j.jpainsymman.2007.07.004

Ernst E (2008b) Complementary and alternative medicine: what the NHS should be funding? Br J Gen Pract 58(548):208–209. doi:10.3399/bjgp08X279562

Ernst E, Pittler MH, Wider B et al (2008) Oxford handbook of complementary medicine. Oxford University Press, Oxford

Gliedt JA, Hawk C, Anderson M et al (2015) Chiropractic identity, role and future: a survey of North American chiropractic students. Chiropr Man Therap 23(1):4-014-0048-1. eCollection 2015. doi:10.1186/s12998-014-0048-1

Hougham P, Parrott A (2017) Holistic medicine and holistic education. British Acupuncture Accreditation Board. http://baab.co.uk/articles/item/14-holistic-medicine-and-holistic-education.html. Accessed 9 Mar 2017

MacLennan AH, Morrison RGB (2012) Tertiary education institutions should not offer pseudoscientific medical courses. Med J Aust 196(4):225–226. doi:10.5694/mja12.10128

Marcus DM, McCullough L (2009) An evaluation of the evidence in "evidence-based" integrative medicine programs. Acad Med 84(9):1229–1234. doi:10.1097/ACM.0b013e3181b185f4

Sampson W (2001) The need for educational reform in teaching about alternative therapies. Acad Med 76(3):248–250. doi:10.1097/00001888-200103000-00011

School of Homeopathy (2017) A holistic view of health. www.homeopathyschool.com/why-study-with-us/what-is-homeopathy/holistic-view-of-health/. Accessed 12 Mar 2017

Scrutton S (2017) Measles. Why homeopathy? http://s-scrutton.co.uk/Why_Homeopathy/illness-i-m/measles.html. Accessed 10 Mar 2017

Smith KR (2011) Factors influencing the inclusion of complementary and alternative medicine (CAM) in undergraduate medical education. BMJ Open 1(1):e000074. doi:10.1136/bmjopen-2011-000074

UCAS (2017) At the heart of connecting people to higher education. UCAS. https://www.ucas.com/. Accessed 16 Mar 2017

White B (2017) Making evidence-based medicine doable in everyday practice—family practice management. Family Pract Manag. http://www.aafp.org/fpm/2004/0200/p51.html. Accessed 10 Mar 2017

Chapter 5
Informed Consent

Informed consent is not just a basic ethical principle, it is a precondition for any medical or surgical procedure or diagnostic test. This means it is not a choice: it is mandatory in all areas of healthcare. Yet, in CAM, informed consent is often neglected. A survey of UK chiropractors, for instance, showed that only 23% always discuss serious risks of their treatments with their patients (Langworthy and le Fleming 2005).

Essentially, there are 4 facets of informed consent:

  i. all relevant information must be disclosed to the patient;
 ii. the patient must fully understand what they have been told;
iii. the patient's decision must be free from coercion or manipulation;
 iv. the patient must be competent, i.e. they must possess decision-making capacity.

To render this discussion less theoretical, let us first consider three scenarios involving a virtual case of an asthma patient consulting a chiropractor. We have chosen chiropractic merely as an example—the issues outlined below apply to most other forms of CAM as well.

SCENARIO 1

Our patient is experiencing breathing problems, and a neighbor told him that chiropractors can help this kind of condition. He consults a 'straight' chiropractor, i.e. a clinician who adheres to the notions of 'subluxation', as taught by the founding father of chiropractic, DD Palmer, about 120 years ago. She explains to the patient that chiropractors use a holistic approach. By adjusting subluxations in the spine, she is confident to stimulate a healing process which will naturally ease the patient's breathing problems. No conventional diagnosis is discussed, nor is there any mention of the prognosis, likelihood of benefit, risks of treatment or alternative therapeutic options.

© Springer International Publishing AG 2018
E. Ernst and K. Smith, *More Harm than Good?*
https://doi.org/10.1007/978-3-319-69941-7_5

SCENARIO 2

Our patient consults a chiropractor who does not believe in the 'subluxation' theory of chiropractic. She conducts a physical examination of our patient's spine and diagnoses several spinal segments that are 'blocked'. She tells our patient that he might be suffering from asthma and that spinal manipulation might remove the blockages and thus increase the mobility of the spine. This, in turn, would alleviate his breathing problems. She does not mention risks of the proposed interventions nor other therapeutic options.

SCENARIO 3

Our patient visits a chiropractor who considers herself a back-pain specialist. She takes a medical history and conducts a physical examination. Subsequently she informs the patient that his breathing problems could be due to asthma and that she is neither qualified nor equipped to ascertain this diagnosis. She tells our patient that chiropractic is not an effective treatment for asthma but that his GP would be able to firstly make a proper diagnosis and secondly prescribe an effective treatment for his condition. She writes a brief note summarizing her thoughts and hands it to our patient to give it to his GP.

As we will explain in this chapter, in scenario 1 and 2, the chiropractor failed the requirements of informed consent. Only scenario 3 describes behavior that is ethically acceptable. But how likely is scenario 3? We fear that it is a rare turn of events. Even if well-versed in both medical ethics and scientific evidence, many if not most CAM practitioners would think twice about providing all the information required for informed consent—because, as scenario 3 demonstrates, providing all the relevant information prevents the patient from receiving any CAM treatment. In other words, CAM practitioners have a powerful incentive preventing them from adhering to the rules of informed consent.

What precisely is meant by providing 'all relevant information' necessary for informed consent? There is general agreement in healthcare that this should include the following elements:

- the indication (or condition) to be treated,
- the nature of the procedure,
- its potential benefits,
- its risks,
- options other than the proposed procedure, including the option of doing nothing at all.

In the remainder of this chapter we shall consider each of these elements in turn.

1 The Indication

Which conditions do CAM practitioners treat? It may seem unrealistic to suppose that, in the above example, some chiropractors would treat asthma. Yet perusal of the Internet readily reveals that many chiropractors indeed claim to be able to treat this condition. For example, a Canadian survey revealed that 38% of chiropractic clinic websites advertised diagnosis, treatment or efficacy for asthma (Murdoch et al. 2016).[1]

This fact can be understood when we consider that DD Palmer taught that 95% of all diseases are caused by subluxations of the spine. This fact explains why many chiropractors see their treatment as a 'cure all'. So, we might think that chiropractors today mainly treat back problems, but in routine practice they treat much, much more. A survey of Australian chiropractors, for instance, showed that, while around half of patients were looking for help with back and neck symptoms, treatment was sought for many other problems besides (Charity et al. 2016). This included various non-spinal musculoskeletal complaints (such as shoulder, hip, knee and leg problems), nonspecific muscle complaints, and headaches. Notably, over a third of patients had no specific symptoms or complaints but sought chiropractic care for the maintenance of good health.

For which of these indications is there enough evidence to justify chiropractic interventions? Here is a summary of our assessment of the evidence:

- back and neck pain: an optimist might grant that there is some promising (but by no means conclusive) evidence.
- shoulder, hip, knee and leg problems, and nonspecific muscle complaints: no good evidence.
- headache: some promising (but by no means conclusive) evidence.
- health maintenance: no good evidence.

In other words, most of the conditions chiropractors treat are not backed up by sound evidence.

This conclusion is supported by another Australian survey which collected data from over 4000 chiropractor-patient encounters (French et al. 2013). The most frequent care provided by the chiropractors was spinal manipulative therapy and massage. Back problems were managed at a rate of 62 per 100 encounters. Musculoskeletal problems comprised 60 per 100 encounters, while 39 per 100 encounters dealt with health maintenance. Other problems managed included headaches, nerve-related problems and depression. These findings confirm that most chiropractors treat non-spinal conditions for which there is no evidence that their interventions are effective.

Perhaps the most implausible notion paraded by chiropractors is the use of chiropractic manipulation for 'health maintenance' or 'wellness' in healthy

---

[1]Chiropractors are not alone: the Canadian survey also revealed that most naturopaths, acupuncturists and homeopaths offer to diagnose or treat asthma.

individuals. As indicated by the above two surveys, the belief that 'chiropractic care' can serve as a form of preventative medicine is widely held amongst chiropractors and patients. Chiropractors frequently recommend that healthy individuals, including children and infants, receive regular chiropractic checkups and maintenance therapy—all for a fee, of course. Yet there is a distinct lack of reliable evidence showing chiropractic maintenance care to be effective (Ernst 2009). In other words, chiropractors earn their living mostly by being economical with the truth regarding the lack of evidence for their actions.

It seems that not just chiropractors but many, if not most, CAM therapists systematically misinform their patients about the value of their interventions for specific conditions. One of us [Ernst] has investigated this issue in some detail. Here are the verbatim conclusions of some of these investigations:

The most popular websites on CAM for cancer offer information of extremely variable quality. Many endorse unproven therapies and some are outright dangerous (Schmidt and Ernst 2004).

Many CAM providers have a negative attitude towards immunization and means of changing this should be considered (Schmidt and Ernst 2003c).

Most of the advice about herbal medicine available via the Internet is misleading at best and dangerous at worst. Potential Internet users should be made aware of these problems and ways of minimizing the risk should be found (Ernst and Schmidt 2002).

Most chiropractors and their professional associations make therapeutic claims that are not supported by sound evidence. (Ernst and Gilbey 2010)

Advice given by CAM practitioners to diabetic patients had the potential to kill them (Schmidt and Ernst 2003b).

The advice chiropractors provide to asthma patients is not based on evidence and has the potential to seriously harm patients (Schmidt and Ernst 2003a).

Anthroposophic doctors often advise against measles vaccinations and can thus be responsible for measles outbreaks (Ernst 2011).

The problem of misinforming potential patients about the conditions to be treated with CAM seems particularly obvious with homeopathy. To take just one (typical) example, an online homeopathy magazine—a professionally presented offering clearly aimed at potential patients—advocates homeopathy for almost every conceivable condition (Homeopathy Information 2017). Under the heading *"illnesses that respond best to homeopathy"*, the magazine lists no less than 12 disparate categories of disorder, ranging from behavioural and psychiatric disorders to urological disorders, as follows:

behavioural and psychiatric disorders, cardiovascular problems, dermatological problems, digestive problems, endocrine disorders, ENT and bronchial problems, gynaecological problems, neurological disorders, osteoarticular complaints, pediatric problems, traumas, urological disorders.

Within these categories, over 80 individual disorders or subcategories of disorder are specified, ranging from asthma to obesity to varicose veins, as follows:

acidity, acne rosacea, acne vulgaris, ankle pain, anxiety, arthritis, arthrosis, asthma, back pain, behavioural issues, bladder etc., blepharitis, bone fractures etc., bronchial tubes, bronchitis, canker sores, conjunctivitis, connective tissue, constipation, contractures etc., contusions, dacryocistitis, depleted immune defences, depression, diarrhoea, duodenal ulcer, ear infections, eczema, educational attainment issues, elbow pain, eye problems, falls, flatulence, fullness, headaches and migraines, heartburn, herpes simplex and zoster, high blood pressure, hives, hyperthyroidism, hypothyroidism, infertility, knee pain, knocks, larynx, leg heaviness, lungs, menopausal complaints, mental fatigue, molluscum conta-giosum, nausea, neck pain, obesity, period disorders, period pains, peripheral arterial dis-ease, pharyngitis, plantar warts, PMT, poor digestion, prostatism, psoriasis, recurrent boils, recurrent infections affecting the throat, recurrent urinary infections, rhinitis, sciatica, shoulder pain, sinusitis, skin complaints, skin, sprains, stomach complaints, stress, styes, teething problems, tonsillitis, tracheitis, trauma of all types, uveitis, varicose veins, venous problems, verucas, vomiting, wrist pain.

This vast list is followed by an astounding statement: "*These are just a few examples, but the list could be endless.*" And beyond the boundless range of conditions treatable with homeopathy, potential patients learn that homeopathy can be used for palliative care, and also for avoiding illness in the first place (because "*homeopathy is an excellent preventive medicine*").

Acupuncture is another modality that claims to be able to treat a huge range of disparate conditions. For example, the website of the British Acupuncture Council (BAC) lists 67 diverse indications that it claims may be treated with acupuncture, as follows (BAC 2017b):

acne, allergic rhinitis, anxiety, arrhythmias and heart failure, asthma, back pain, bell's palsy, cancer care, carpal tunnel syndrome, childbirth, chronic fatigue syndrome, chronic pain, colds and flu, COPD, coronary heart disease, cystitis, dementia, dentistry, depression, dysmenorrhoea, eczema and psoriasis, endometriosis, facial pain, female fertility, fibromyalgia, frozen shoulder, gastrointestinal tract disorders, gout, headache, herpes, HIV infection, hypertension, infertility art, insomnia, irritable bowel syndrome (IBS), kidney stones, male infertility, menopausal symptoms, migraines, multiple sclerosis, nausea and vomiting, neck pain, neuropathic pain, obesity, obstetrics, osteoarthritis, palliative care, Parkinson's disease, PCOS, post-operative pain, post-traumatic stress disorder, premen-strual syndrome, puerperium, Raynaud's, rheumatoid arthritis, sciatica, sinusitis, sports injuries, stress, stroke, substance misuse, tennis elbow, thyroid disease, tinnitus, type-2 diabetes, urinary incontinence, vertigo.

The effects of acupuncture claimed by the BAC range from symptomatic benefits to specific improvements in disease pathology. Symptomatic effects generally refer to a reduction in pain; for example, in the case of dentistry, acupuncture is claimed to relieve dental pain by inter alia leading to the "*release of endorphins and other neurohumoral factors*". Endometriosis provides an example of the latter type of claim, where acupuncture's supposed influence on the diseased reproductive tract include "*reducing inflammation – by promoting release of vascular and immunomodulatory factors*". Another example is Parkinson's disease, where acupuncture is claimed to specifically help in the management of the disease by improving various aspects of brain pathology, including "*attenuating neuronal damage and increasing the number of neurons in the substantia nigra*".

In some cases, the BAC gives caveats where the research supporting these claims is particularly weak. However, by nevertheless listing conditions for which the evidence that acupuncture works is extremely flimsy, contradictory or non-existent, the impression given to potential patients is clear: 'it may work for your disorder'. In any case, most such conditions listed by the BAC are accompanied by the following words:

> In general, acupuncture is believed to stimulate the nervous system and cause the release of neurochemical messenger molecules. The resulting biochemical changes influence the body's homeostatic mechanisms, thus promoting physical and emotional well-being.

This catchall statement implies that a scientific mechanism has been elucidated to explain how acupuncture may help patients with these varied conditions. While potential patients may be reassured by these words, this statement is nevertheless vague and scientifically unsubstantiated. At best, it may be interpreted as alluding to the operation of a placebo effect in the brain. If this is its meaning, why couch it in such opaque scientific-sounding language? We suspect the answer is: To pretend to patients that acupuncture is more than merely a placebo.

Whether one looks at acupuncture, chiropractic, homeopathy, or indeed any other CAM modality, the overall picture is depressingly clear: each type of CAM claims to be able to treat a very wide range of medical conditions. This 'one size fits all' ideology is a central feature of the CAM proponents' mind-set. In many cases this is because, as discussed in the previous chapter, individual CAM modalities usually posit a single universal phenomenon (such as blockage of Qi, or subluxation of the vertebrae) as the underlying cause of disease. But these interlinked concepts of 'one cause' and 'one treatment' are completely incompatible with medical science: in short, they are pure fantasy.

Related to the nonsensical 'one cause' explanation of the origin of disease, CAM practitioners claim to treat the *root causes* of disease. A traditional acupuncturist, for instance, might be convinced that all disease is the expression of an imbalance of life-forces, and that needling acupuncture points along meridians will restore the flow of Qi and re-balance yin and yang, thus restoring health. He therefore would automatically assume that he is treating the root causes of any condition. Similarly, if a chiropractor believes that all diseases are due to 'subluxations' of the spine, it will seem logical to him that spinal 'adjustment' is synonymous with treating the root cause of whatever complaint his patient is suffering from. Or, if a Bowen therapist is convinced that "*the Bowen Technique aims to balance the whole person, not just the symptoms*", he is bound to be equally sure that "*practically any problem can potentially be addressed*" by this intervention (Cavanagh 2017). These are just three examples: the idea that diseases have a single root cause is central to most CAM modalities. (Each form of CAM claims its own version of causality, and one is bound to say: they cannot all be right!) Any successful treatment of a root cause means that the therapy in question completely heals the problem at hand. If we abolish the cause of a disease, we would expect the disease to disappear for good.

By contrast, conventional medicine is not limited by single root cause doctrine; medical scientists research the cause of each disease free from the ideological blinkers restricting the scope of imagination of their CAM counterparts. It is not a coincidence that, while we have no difficulty naming diseases which conventional medicine can cure, we cannot name a single alternative treatment for which there is compelling evidence proving that it can produce more than symptom-relief. Crucially, CAM practitioners cannot name such a scenario either. In other words, the notion that CAM tackles the root causes of disease is a myth and an unethical distraction from the truth.

1.1   Failure to Obtain True Informed Consent: Ethical Implications

The information on indications supposedly treatable by CAM set out above gives a broad picture of the misinformation around CAM which consumers are exposed to. In the context of individual patients consulting CAM practitioners, informed consent requires that the patient must understand why a particular treatment is being offered. In turn, this requires that the patient receives a valid explanation of the mechanism by which the treatment will operate, in relation to the underlying cause of the disorder, which must also be accurately explained to the patient.

In conventional medicine, such explanations are usually possible. For example, in the case of antibiotics used to treat an infection, the patient can be advised that microorganisms have colonised her body, and that the antibiotic will attack and destroy the microorganisms. All of this is based on sound scientific theory and a wealth of empirical evidence. In the case of other conventional treatments, the explanation may have to be a qualified one. For example, a novel cancer drug may not have had its full mechanistic effects determined, and effectiveness and safety may remain to be fully established. Nevertheless, because the putative mechanism of action of the drug is consistent with well-founded scientific knowledge and principles, these limitations can be explained to the patient such that a rational decision on whether to consent can be made.

By contrast, most CAM modalities are based upon irrational ideas about the causes of disease. Even where an internally consistent mechanistic framework has been posited, this cannot be sufficient as a source of valid explanation, if it is incompatible with established science. Consequently, CAM practitioners cannot normally provide patients with valid explanations for their diagnoses and treatments. This leads to a stark and far-reaching conclusion: informed consent is simply not possible for these CAM modalities (Shahvisi 2016). Because informed consent is a cornerstone of medical ethics, CAM practitioners who diagnose and treat patients in the absence of genuine informed consent are guilty of unethical behaviour. Because the majority of CAM modalities are based upon unscientific and irrational concepts, this ethically unacceptable failure to obtain true informed consent is almost ubiquitous amongst CAM practitioners. This is a fundamental, serious and widespread ethical problem for CAM.

1.2 Diagnostic Techniques and CAM

Most people think of CAM as a range of therapies, and tend to forget that CAM practitioners also use of a wide range of diagnostic procedures. Just as informed consent is an ethical requirement for treating a patient, diagnostic procedures equally require patient consent. The ethical principle of patient autonomy is universally applicable, and covers diagnosis as well as treatment. Additionally, the ethical principle of non-maleficence provides further justification: many diagnostic procedures have a potential to harm the patient. This, of course, applies to any type of medicine and includes CAM. Therefore, patients need to know about the risks and must be free to accept or decline diagnostic procedures.[2] Regrettably, many CAM practitioners employ potentially harmful diagnostic tools in the absence of meaningful patient consent.

Chiropractors, for instance, have been shown to over-use X-rays, a practice which exposes their patients not only to unnecessary costs but also to unjustifiable carcinogenic risk (Ernst 1998). A study from the US has demonstrated that the rate of spine radiographs within 5 days of an initial patient visit to a chiropractor is 204 per 1000 new patient examinations (Bussieres et al. 2013). Patients with non-specific back pain (by far the most common reason for consulting a chiropractor) do not normally need X-rays. Therefore, such rates are far too high and expose patients to avoidable risks in the absence of any benefits. Patients do not usually question the use of X-rays. But how many patients would agree to these X-rays, if the chiropractor provided an honest and accurate explanation of risk-versus-benefit, outlining that they are firstly unnecessary and secondly not free of risk?

The purpose of a diagnostic test is, of course, to establish the presence or absence of an abnormality, condition or disease. Conventional doctors use all sorts of validated diagnostic methods, from physical examinations to laboratory tests, from blood pressure measurements to imaging techniques. By contrast, CAM practitioners use mostly alternative methods for arriving at their diagnoses. Acupuncturists, iridologists, spiritual healers, massage therapists, reflexologists, applied kinesiologists, homeopaths, chiropractors, osteopaths and many other types of CAM practitioners all have their very own diagnostic techniques which are

---

[2]In conventional medicine, it is usual to ask for *explicit* (often written) consent where a diagnostic procedure carries a significant risk of serious harm to the patient (e.g. exploratory surgery under general anaesthesia); by contrast, *implied* consent is generally deemed acceptable for low risk, non-invasive procedures (e.g. taking a sample of venous blood). In the latter case, the patient's behaviour (i.e. not refusing the procedure) is considered to be sufficient evidence of consent. But in all cases, it is necessary that the patient has been provided with a sufficient explanation of the diagnostic procedure, otherwise consent will not be 'informed'.

unknown in conventional healthcare. This obviously begs the question: how reliable are these methods? Even if a given diagnostic approach is safe, if it is incapable of reliably yielding accurate diagnostic information, it is ethically fraught because it is likely to result in misdiagnosis and unnecessary treatments as well as costs.

Unfortunately, there is little reliable information on the validity of alternative diagnostic approaches; it is very rare for a CAM diagnostic procedure to be subjected to rigorous scientific testing (Ernst et al. 2006). In the few cases where such tests have been conducted, the results generally do not encourage confidence. But none of this prevents CAM practitioners from continuing to use unproven and even disproven diagnostic methods.

For example, most chiropractors use the Kemp's test, a manual procedure intended to diagnose problems with lumbar facet joints. The chiropractor rotates the torso of the patient, while her pelvis is fixed; if this procedure causes pain, it is interpreted as a sign of lumbar facet joint dysfunction which, in turn, would be treated with spinal manipulation. A systematic review evaluated the existing literature regarding the accuracy of the Kemp's test in the diagnosis of facet joint pain compared to a reference standard (Stuber et al. 2014). All diagnostic accuracy studies comparing the Kemp's test with an acceptable reference standard (totaling five studies) were included in the review, with the studies being scored for quality and internal validity. The authors concluded: "*currently, the literature supporting the use of the Kemp's test is limited and indicates that it has poor diagnostic accuracy. It is debatable whether clinicians should continue to use this test to diagnose facet joint pain.*"

This suggests that chiropractors frequently treat conditions which the patient does not have. This, in turn, is not just a waste of money and time but also, if the ensuing treatment is associated with risks (which chiropractic spinal manipulation clearly is), an unnecessary exposure of patients to harm. And it needs to be asked: were the patients on which Kemp's test was used given a full and accurate explanation of the test, including a statement on the lack of supportive evidence? If so, it would be surprising that rational patients nevertheless decided to undergo Kemp's test. An alternative explanation is that the chiropractors, wedded to dogma-driven techniques and seemingly impervious to scientific evidence, provided their patients with a biased, untruthful account of the test (that is, if they discussed it at all). If this is so, then informed consent has been lacking—which is, of course, a violation of medical ethics.

Some branches of CAM are *purely* diagnostic, and do not offer treatments per se. Take for instance iridology. Ignaz von Peczely (1826–1911), a Hungarian physician, got the idea for iridology (or iris-diagnosis) more than a century ago, after seeing streaks in the iris of a man he was treating for a broken leg, and similar phenomena in the iris of an owl whose leg von Peczely had broken many years before. He subsequently became convinced that the study of the iris could distinguish between healthy organs and those that are overactive, inflamed, or distressed. We hardly need point out that Peczely's logic was highly flawed, and his resultant conclusions frankly preposterous.

Despite its inauspicious origins, iridology is practiced to this day. Modern iridologists claim to be able to diagnose the health status of an individual, medical conditions and predispositions to disease through supposed abnormalities of pigmentation in the iris. The modality became internationally known when US chiropractors began adopting it in their clinical practice.[3] In the United States, most insurance programs do not cover iridology but, in European countries, they often do. This is the case in Germany, for instance, where the majority of the Heilpraktiker (non-medically qualified health practitioners) practice iridology (Mueller 2016).

The popularity of iridology renders it necessary to ask whether, despite its implausibility, the method might nonetheless have some empirical validity. One of us investigated this some years ago and published a systematic review, in which all then available tests of iridology as a diagnostic tool were critically evaluated (Ernst 1999). Four case control studies were included; these are investigations where iridologists are asked to tell by looking at the iris of individuals whether that person does or does not have a certain condition. Most of these studies suggested that iridology is not a valid diagnostic method. The systematic review concluded: *"the validity of iridology as a diagnostic tool is not supported by scientific evaluations. Patients and therapists should be discouraged from using this method."* No persuasive evidence has emerged since the publication of Ernst's paper to challenge this conclusion.

Because unreliable diagnostic tests can generate both false positive and false negative results, the danger of using them is not just academic but all too real. If the method generates a false positive result, an alert will be issued in vain, people will get anxious for nothing, unnecessary treatments will be administered, and time and money will be lost. If the method generates a false negative result, patients will assume they are safe while, in fact, they are not. In extreme cases, such an error will cost lives. In this way, invalid diagnostic tests are akin to bogus bomb-detectors.

To us, the following conclusion seems incontrovertible: CAM practitioners who choose to employ disproven, unproven or unreliable diagnostic methods are guilty of ethically reprehensible practice. Because they frequently use these procedures without obtaining informed consent (otherwise patients would object to being submitted to and paying for them), they are also violating a most important ethical principle.

---

[3]One may ask: why would one group of CAM practitioners (in this case chiropractors) adopt diagnostic methods from a fundamentally different CAM modality? After all, there is no obvious logical link between chiropractic conceptions of disease (i.e. subluxations, blockage of spinal segments, etc.) and the notion that iris colouration reflects bodily components. The answer, we believe, lies in the irrational mind-set of the CAM practitioner: freed from the normal imperatives of requiring plausibility and evidence, it becomes easy to uncritically accept superficially appealing yet fundamentally defective medical notions.

2   The Nature of the Procedure

The fact that real consent is often impossible for CAM treatments might rule out many CAM approaches on ethical grounds. Nevertheless, the reality is that CAM practitioners do, of course, continue to treat patients, apparently untroubled by ethical concerns around informed consent. One way that CAM proponents try to justify themselves in this context concerns the nature of the procedures offered. The argument here is that, despite most CAM therapies lacking a valid explanatory rationale, it is nevertheless permissible to use them on grounds of one or more of the following claims:

- CAM is natural and therefore safe,
- CAM is individualised, and therefore defies scientific validation,
- CAM does not work as a conventional therapy, rather it helps to 'detox' your body,
- CAM treatments have stood the test of time and are therefore valid.

These are common claims made by many, if not most CAM proponents. The general idea here is that, while there is no scientific basis for a therapy and no robust clinical evidence exists to show that the therapy works and is safe, one or more of the above features of CAM nevertheless justifies its use. Patients would therefore be giving their consent to be treated based on such special pleading.

Aside from the question of whether CAM therapists are generally disposed to providing such honest assessments of their creed (the evidence strongly suggests that they are not), the validity of these special pleas needs to be examined.

2.1   CAM: Natural and Safe?

The claim that all CAM therapies are natural and therefore inherently safe is widespread but demonstrably wrong. Implicit in this idea is the notion that conventional medicine is somehow inherently unnatural, relying heavily on harmful synthetic chemicals. Nature, by contrast, is pictured as benign, and natural remedies are therefore not just intrinsically superior but also safer. While undoubtedly effective for marketing purposes, this argument is nevertheless false and in many instances outright dangerous.

Not all forms of CAM are natural. For instance, there is nothing natural in sticking needles into a patient's body (as in acupuncture), or endlessly diluting and shaking a medicine (as in homoeopathy), or in introducing liters of tepid coffee to the body via the rectum (as in the coffee enemas used by some CAM practitioners).

Moreover, nature is by no means always benevolent, as anyone who has been out at sea in heavy weather or had the misfortune to be hit by lightning knows all too well. Even 'natural' plant extracts are not necessarily natural, as they can contain active ingredients in much higher concentrations than nature intended. And they are certainly not all safe—just think of hemlock.

2.2 Individualized CAM Treatment

It is often claimed that the individualized nature of CAM means that the usual standards of evidence-based medicine do not apply. If this were true, then the stringency of the ethical requirement for informed consent would be weakened.

Herbal medicine serves as a good example on which to examine claims that the individualized nature of CAM excuses it from the normal requirements of informed consent. We are made to believe that herbal medicine is one entity, yet there are two dramatically different categories of herbal medicine, and the proper distinction of the two is crucially important (Ernst 2007). The first type essentially uses reasonably standardized and well-tested herbal remedies to treat specific conditions. This has a plausible basis: many plants contain substances that can act strongly in the human body (for good or ill), and many mainstream drugs are derived from botanicals. For example, St John's wort has shown some effectiveness as a treatment of depression (Maher et al. 2016). This approach has been called by some experts 'rational phytotherapy' (Schulz 2011).

The second type of herbal medicine, generally known as 'traditional herbalism', entails consulting an herbal practitioner who will take a history, make a diagnosis according to obsolete concepts and prescribe a mixture of several herbal remedies tailor-made to the characteristics of his patient. Thus 10 patients with the identical diagnosis (say depression) might receive 10 different mixtures of herbs. This is true for traditional herbalism of all traditions, e.g. Chinese, Japanese, Tibetan, Indian or European, and virtually every traditional herbalist you might consult will employ this individualized approach.

Many consumers know that, in principle, there is some reasonably good evidence for several specific herbal medicines such as the above-mentioned St John's Wort. They fail to appreciate, however, that this does *only* apply to what we have just called 'rational phytotherapy'. Consequently, consumers may consult herbalists in the belief that they are about to receive an evidence-based therapy. Nothing could be further from the truth! The individualized approach is not evidence-based; even if the individual extracts employed were all supported by sound data (which they usually are not) the mixtures applied are clearly not. Several studies testing individualized herbalism have been published and their results fail to confirm that such treatments are effective for any condition (Guo et al. 2007).

From the perspective of informed consent, rational phytotherapy is in principle acceptable, because true patient understanding is often possible, even though knowledge of the biological mechanisms through which the medicine acts may be limited. In this respect, rational phytotherapy is not dissimilar to the above-mentioned example of consent being obtained for a novel cancer drug. By contrast, the second form of herbal medicine is not rationally explicable to patients. Unless it is considered acceptable under the aegis of informed consent to tell the patient 'this potion will heal you in magical ways', informed consent cannot be obtained for this type of herbalism. The fact that the treatments emanating from traditional herbalists are individualized to each patient does not provide an ethical get-out from the need to obtain informed consent.

2.3 CAM Does Not Work as a Conventional Therapy, Rather It Helps to 'Detox' Your Body

Many forms of CAM implicitly or explicitly include the notion of 'detox'. This term is used in conventional medicine to describe the therapies used for weaning addicts off drugs. In CAM, it is employed differently, and it is this latter purpose that we are addressing here.

In CAM, detox describes any means of ridding the body of toxins, real or imagined. Many CAM practitioners claim that our body is constantly bombarded with a range of poisons which originate from the food we eat, the environment we live in, the medicines we take, the air we breathe, etc. This assumption is, of course, not entirely incorrect. Modern life is never totally free of toxins. (Even a pre-industrial lifestyle would not be devoid of dietary toxins, because plants frequently produce toxic chemicals, as part of their defences against pathogens and insects.) In CAM, the dangers arising from toxins in our bodies are, however, often exaggerated.

The reason for this exaggeration is that CAM practitioners claim to have a solution to the supposedly ubiquitous problem of toxins: the therapy they offer happens to be an effective way of eliminating these dangerous agents. Regardless of whether we listen to naturopaths, homeopaths, herbalists, aromatherapists, colon-therapists etc., we are likely to hear this claim. CAM practitioners employ medicines, ear-candles, manual techniques, foot-baths, diets, colonic irrigation and many other techniques for detoxifying our systems.[4]

The core scientific problem with these therapies is clear: practitioners cannot name the toxins they pretend to eliminate. If they did, we could measure the blood levels before and after the therapy and thus test the effectiveness of their methods. But because tests for unnamed substances cannot be conducted, positive disproof of toxin elimination is unavailable. However, logic dictates that the onus lies with those making these claims to provide supporting evidence. Even though the necessary biochemical tests for known toxins are in principle very straightforward, and ought to yield unequivocal quantitative results, evidence of the supposed detoxification abilities of CAM has been limited to a very small number of published reports, each presenting rather equivocal or questionable results. As with research claims for the effectiveness of CAM treatments of specific disorders, as discussed earlier in this book, isolated positive reports rarely provide compelling evidence, and are more likely to mislead than to inform.

For example, a survey of naturopathic doctors (NDs) in the United States reported that 83% of NDs claimed to use follow-up measurements to determine effectiveness of detoxification therapies (Allen et al. 2011). Aside from patient questionnaires and medical histories, a direct biochemical measurement was

---

[4]A related issue is the notion of conventional medicines as poisons: this mantra is frequently used by CAM proponents to alienate consumers from real healthcare and encourage them into the open arms of quacks.

allegedly used in 53% of cases, namely 'urinary provocative challenge testing'. However, this paper provides no further details as to what this testing entailed. Moreover, the notion of 'provoked testing' has itself been called into question, as its results can be unreliable and potentially misleading to patients (Barrett 2017).

To take another example, a 2000 study examined the use of a short programme of detoxification in disease-free individuals (MacIntosh and Ball 2000). Working on the basis that "*removal of toxins from the body is an integral part of Ayurvedic, yogic, and naturopathic medicine*", and apparently motivated by the (unproven) notion that "*symptoms of poor health in people free from diagnosed disease may be related to toxin buildup*", the researchers subjected 25 volunteers to 7 days of 'detoxification'. This comprised a hypoallergenic diet, 6 scoops of a 'medical food supplement' (UltraClear), and at least 2 quarts of filtered water daily.

The reported biochemical results included a 23% increase in "liver detoxification capacity" (as measured by caffeine clearance from the body) and an increase in the ratio of sulphate:creatine in the urine, indicating "a trend toward improved liver function".

However, the reliability and importance of these results is highly questionable. Many profound weaknesses are evident in this study, but two flaws effectively torpedo its findings: firstly, the trial had no control group, and secondly the biochemical results were not even statistically significant. This study was published over 15 years ago: had the reported detoxification effects been real, one would expect follow-up studies to have been conducted, producing results that built upon these initial findings. That this has not occurred is an incriminatory testament to the (lack of) validity of this isolated study.

In sum, CAM researchers have manifestly failed to produce robust evidence to support their claims for toxin elimination. The simple fact is that none of the treatments advocated for 'detox' have ever been shown to eliminate anything other than a patient's cash. The claim that they do what they are advertised to do is, in other words, false.

In terms of informed consent, detox presents a major problem. If a practitioner told his patients that 'detox' is pure fantasy, they would most likely decline to receive the proffered treatment. It follows that, in the majority of cases, detox is being used under false pretences and in the absence of genuine informed consent. This is ethically unacceptable.

2.4   CAM Treatments Have Stood the Test of Time and Are Therefore Valid

Many forms of CAM have a history of hundreds (homeopathy, chiropractic etc.) or even thousands (acupuncture, herbalism etc.) of years. The notion that a therapy has stood the test of time convinces many that it must be effective and safe; had it been ineffective or harmful, it would simply no longer exist, they think. Certainly, CAM enthusiasts try to make us believe that passing the test of time amounts to strong evidence of effectiveness; some even claim that such evidence is more meaningful than clinical trials or other scientific tests.

The truth, however, is different: while the test of time can sometimes be a valuable indicator, it can never be a proof for anything. In fact, the antiquity of a treatment like acupuncture, for instance, might simply demonstrate that the therapy originates from a time when even the most basic knowledge about anatomy, physiology, pathology etc. was not available. Seen from this perspective, the long history of a treatment is hardly something to commend it, and might merely indicate its relative primitivity and uselessness.

Astrology serves as a useful analogy. The positions of celestial objects have been used for millennia as a source of information about human affairs. In this way, astrology has certainly 'stood the test of time'. However, modern science has debunked these notions: we now know that the positions of planets and stars cannot directly influence human events on earth, and experimental studies have confirmed that astrology is devoid of explanatory power. Thus, astrology is now recognized as pseudoscience, and few educated people nowadays place faith in its supposed predictive powers.

Why has astrology largely been rejected while CAM modalities such as acupuncture remain popular on grounds of having 'stood the test of time'? One major reason is that, while most modern-day citizens have a relatively good understanding of astronomy (after all, details of the solar system are routinely taught to children from elementary school onwards), there exists a far lower level of understanding of physiology amongst the public. This lack of knowledge opens the door to quacks to dupe their patients.

The history of medicine is littered with examples of treatments that had 'stood the test of time', were used abundantly, and only abandoned once scientific tests were conducted and failed to show that they were neither safe nor effective. Take for instance blood-letting, mercury cures, leeches and many more. The impression of effectiveness can be false because factors other than the therapy brings about an improvement of symptoms. And the impression of safety can be mistaken for a range of reasons such as lack of adequate monitoring systems, or delayed adverse effects which are not attributed to the right cause.

3 The Potential Benefits

We have argued above that the supposed mechanisms of action of most CAM modalities are fundamentally inexplicable to patients, thus torpedoing the 'in-formed' part of the consent requirement. We have furthermore argued that neither the 'natural' character of CAM therapies nor the fact that these treatments are often individualized can rescue CAM from this charge. Nevertheless, a permissive argument is conceivable, in which the proven effectiveness of a therapy provides sufficient grounds for informed consent. So, the practitioner would explain to the patient that, although the mechanism of action of the therapy is inexplicable, good evidence exists to show that the therapy nevertheless works. It would then, on this permissive view, be acceptable to invite the patient to consent to the therapy.

We consider this view to be problematic: it seems like another case of special pleading on the part of CAM advocates. However, let us for the sake of argument suppose that it is valid. On that basis, the crucially relevant factors become effectiveness and risk. We shall deal with risks from CAM treatments later in this chapter. In terms of effectiveness, Chapter 3 (The Reality of CAM Research) provided several examples of therapies of doubtful effectiveness. In that context, we stressed that published CAM research tends to be of low quality and often paints a false-positive picture of clinical effectiveness. In the respect of the permissive view of consent, the crucial issue is this: has the prospective patient been presented with accurate information on effectiveness?

One main reason for the current popularity of CAM is the fact that, on the Internet, in books, in newspapers and magazines, CAM therapies are relentlessly being promoted as effective. Yet, when one looks at the basis for such claims, it almost invariably transpires that the evidence is sorely lacking, and that the CAM claims are either false or grossly over-optimistic. Of course, the promulgation of such propaganda occurs mostly outside of the consulting room, and as such—however inherently objectionable such material may be—a violation by the practitioner of clinical ethics does not occur. Nevertheless, media claims for the effectiveness of CAM are likely to be behind the patient's decision to seek CAM in the first place. This makes it all the more crucial that the practitioner presents accurate information on therapeutic effectiveness when seeking the patient's consent.

But many CAM practitioners are themselves frequently involved in the production of media promotions of therapeutic effectiveness, and presumably believe (uncritically) in the truth of these messages. In any case, CAM therapists will almost automatically reassure their patients that what they have read on the Internet or seen on TV is true; otherwise they would be out of business! Yet, as we argued in the last chapter, much of the supposed evidence published in the academic press—and upon which many media claims are based—is so flawed as to be meaningless and misleading. When CAM practitioners base the advice they give to patients on such sources, or fail to contradict misinformation when a patient has assimilated it, genuine informed consent becomes impossible.

Practitioners have an ethical duty to avoid uncritically accepting overly-rosy claims of effectiveness; sadly, many CAM practitioners are frequently failing to satisfy this important requirement. Even where strong evidence of a *lack* of clinical effectiveness exists, many CAM practitioners nevertheless continue to treat patients, and presumably fail to mention (or possibly misrepresent) evidence of ineffectiveness when seeking consent.

This even applies to the widely-accepted belief that chiropractic effectively alleviates chronic low back pain. A recent study performed a randomized placebo-controlled trial comparing the effectiveness of chiropractic spinal manipulative therapy (cSMT) to a sham (placebo) intervention (Dougherty et al. 2014). A total of 136 elderly patients suffering from chronic low back pain (LBP) were included in the study—with 69 being randomly assigned to cSMT and 67 to the sham intervention. Patients were treated twice per week for 4 weeks. The outcomes were assessed at baseline, 5, and 12 weeks post baseline. Both groups demonstrated significant decrease in pain and disability at 5 and 12 weeks. At 12 weeks, there was no significant difference in pain and a statistically significant decline in disability scores in the cSMT group when compared to the control group. There were no significant differences in adverse events between the groups. The authors concluded that cSMT "*did not result in greater improvement in pain when compared to our sham intervention; however, cSMT did demonstrate a slightly greater improvement in disability at 12 weeks. The fact that patients in both groups showed improvements suggests the presence of a nonspecific therapeutic effect*".

Of course, this is just one study of many, and for a more reliable verdict we need to see what the systematic reviews tell us about cSMT and chronic LBP. Fortunately, an authoritative review has been published, based on 26 RCTs; its conclusions are as follows: "*High quality evidence suggests that there is no clinically relevant difference between SMT and other interventions for reducing pain and improving function in patients with chronic low-back pain.*" (Rubinstein et al. 2011).

For most other forms of CAM, the gap between claims made by CAM practitioners and the evidence to support them is even larger. Homeopaths, for instance, frequently quote the so-called 'Swiss report' on homeopathy (Bornhöft and Matthiessen 2011), a pro-homeopathy document allegedly based on an official examination of homeopathy by the Swiss government. Contrary to such claims, this report is neither official nor from a government; its flawed nature had been disclosed over and over again (Ernst 2017).

Other homeopaths promote homeopathic remedies as a cure for serious diseases including AIDS (Onlymyhealth 2017) and cancer (Der-Ohanian 2008). Yet, as we have pointed out previously, the truth about homeopathy has been known for some time: homeopathy is an utterly implausible modality, and highly diluted homeopathic remedies are no more effective than placebo.

Perhaps the most thorough investigation in the history of homeopathy was recently conducted by the Australian Government's National Health and Medical Research Council (NHMRC 2017). This investigation unequivocally concluded in the following highly negative terms:

Based on the assessment of the evidence of effectiveness of homeopathy, NHMRC concludes that there are no health conditions for which there is reliable evidence that homeopathy is effective. Homeopathy should not be used to treat health conditions that are chronic, serious, or could become serious. People who choose homeopathy may put their health at risk if they reject or delay treatments for which there is good evidence for safety and effectiveness. People who are considering whether to use homeopathy should first get advice from a registered health practitioner. Those who use homeopathy should tell their health practitioner and should keep taking any prescribed treatments. The National Health and Medical Research Council expects that the Australian public will be offered treatments and therapies based on the best available evidence.

To close our discussion of the ethics of presenting overly-optimistic accounts of CAM effectiveness, we offer the following conclusion: the authors of websites, books and other media that promote implausible, unproven or disproven medical treatments are guilty of ethically reprehensible behaviour. Such falsehoods will frequently lead patients to the door of quacks who are more than happy to support such misleading claims and prescribe ineffective treatments for their patients. These practitioners are in breach of the ethical duty to obtain informed consent, because consent obtained based on misinformation cannot be truly 'informed'.

4   The Risks

Where a CAM practitioner presents a patient with an unduly positive picture of the likely benefits of a treatment, an ethical breach has occurred. A parallel ethical problem emerges when the practitioner misrepresents the risks of a treatment. In both cases, informed consent cannot be said to have been obtained, as the patient has not been truthfully informed. The first type of practitioner behaviour may lead to the patient failing to obtain effective treatment, resulting in indirect harm to their health. By contrast, misrepresenting the risks associated with a treatment threatens to inflict direct harm to the patient. Direct harm is not necessarily of greater ethical magnitude than indirect harm: for example, a homeopath's failure to properly diagnose and treat a major medical problem, such as a malignant tumour, is likely to harm the patient much more than the infliction of a relatively small direct harm, such as a minor skin infection caused by acupuncture. Nevertheless, the accurate description of medical risks is central to the notion of informed consent.

If we believe its proponents, CAM ought to be almost entirely free of risks. This is also the message put out daily by many health writers and many others promoting CAM via books, the Internet, newspapers or other media. Accordingly, patients who seek CAM will often do so on the assumption that their treatment will be 'gentle' and safe in comparison with mainstream treatments. Indeed, a systematic review assessing the reasons why people elect to use CAM identified the hope for fewer adverse effects as one of the most prominent motivations (Ernst and Hung 2011). Sadly, the notion of CAM being risk-free is far from the truth.

Most forms of CAM can generate adverse effects; in some cases, these are serious, even fatal. Currently, we know relatively little about the risks of CAM. The main reason for this void is that there are no effective systems in place to monitor its risks. In conventional medicine, we have post-marketing surveillance systems which, even though far from perfect, will alert us to specific risks and enable us, for instance, to withdraw a dangerous drug from the market. Nothing comparable exists for the vast majority of CAM treatments. Essentially, that means we have to rely on the occasional case report to fill the huge gaps in our knowledge. It seems obvious that this situation must generate an overall picture which is imprecise and invariably under-estimates the real risks.

Despite this significant deficit in reliable information, we do know about some of the risks of CAM. It is beyond the scope of this book to present an exhaustive catalogue of the risks and harms that have been documented for the myriad of CAM modalities. Instead, the following sections will consider just two popular CAM systems, namely acupuncture and chiropractic. We will present a few real-life examples of harm from these modalities, together with supporting academic evidence, to illustrate the issues that apply not only to these two modalities, but to CAM in general.

4.1   Acupuncture

The British Acupuncture Council helps potential patients to understand the risks of acupuncture using the following Q&A:

> Are there any side effects?
>
> Occasionally a small bruise can appear at a needle site. Sometimes people can feel dizzy or tired after a treatment but this passes quickly.

(BAC 2017a)

This is very typical of the way that risks are dealt with by acupuncture proponents and practitioners: very little tends to be said, and what is said usually refers merely to very minor side effects. Patients are thus led to believe that acupuncture carries no serious dangers. But is this true? The short answer to this question is no.

For example, a recent case report told the story of a male patient in his late forties who died immediately after an acupuncture treatment (Watanabe et al. 2015). An autopsy disclosed severe haemorrhaging around the vagus nerve in the neck. When stimulated, the vagus nerve reduces the activity of the heart, and the autopsy found that the man most probably died from a form of heart failure caused by acupuncture-induced stimulation of his vagus nerve.

Death or serious injury following vagus nerve stimulation by acupuncture is a rare occurrence, but acupuncture can harm patients in several other ways. For example, acupuncture of the chest can cause a collapsed lung, or 'pneumothorax'. Indeed, this is by far the most frequently reported serious complication caused by acupuncture; well over 100 instances have been described in the medical literature which, of course, reflects only a fraction of the true (but unknowable) number, due to the chronic underreporting of such events. Nevertheless, new cases are being reported almost on a monthly basis (Hampton et al. 2014).

Other documented serious injuries or deaths from acupuncture include cardiac tamponade, where an acupuncture needle causes a haemorrhage close to the heart. Accumulating blood places physical pressure on the exterior of the heart, preventing its proper function, with often fatal consequences. A 2011 review of all known cases of cardiac tamponade after acupuncture included a total of 26 such incidences (Ernst and Zhang 2011). In 14 patients, the complications were fatal. In addition to cardiac tamponade, hemorrhages following acupuncture damaging other vascular sites can cause a range of serious adverse effects, as evidenced by a number of documented case studies (Bergqvist 2013).

Acupuncture has also been shown to harm patients by transmitting infections. One review from Egypt noted that acupuncture presented a significant risk for acquiring hepatitis C infections (El-Ghitany et al. 2015). Another report described the diagnosis, treatment and follow-up of 30 patients presenting with acupuncture-induced primary inoculation tuberculosis (Liu et al. 2014).

The medical literature also contains a variety of case studies in which individual patients have sustained serious harm due to acupuncture needles damaging one or other internal body organ. This demonstrates that acupuncture needles have the potential to inflict serious damage wherever used on the body. This is, of course, quite a different reality from the abovementioned acupuncturists' attempt to reassure patients that the adverse effects of acupuncture are restricted to the 'occasional small bruise'.

In total, around 100 deaths have been reported after acupuncture in the medical literature (Ernst 2010a). A recent study found that the incidence of any adverse events per patient was 42.4% with traditional acupuncture, 40.7% with minimal acupuncture and 16.7% with non-invasive sham acupuncture (Chung et al. 2014). These figures are much higher than those previously reported, which were around 10% (White et al. 2001).

When presented with such findings, acupuncture fans almost inevitably respond that the number of reports of fatal or serious adverse effects is trifling. It is true that the reported numbers are small, but they could well be just the tip of a much larger iceberg: there is no reporting system in place to pick up all severe complications, and in the absence of such a scheme, nobody can provide reliable incidence rates. And even if the numbers of severe complications and deaths were small, even a single fatality would seem one too many, considering the paucity of good evidence existing to suggest that acupuncture is more effective than placebo treatment.

4.2 Chiropractic

The British Chiropractic Association (BCA) states:

> Chiropractic treatment mainly involves safe, often gentle, specific spinal manipulation to free joints in the spine or other areas of the body that are not moving properly.

(BCA 2017)

Presenting chiropractic as 'safe' is typical for the chiropractic profession, and patients hearing such words will assume that the risks of receiving chiropractic treatment must be very low. Sadly, this assumption is not well founded, as evidenced by a (growing) number of reports of serious adverse effects from chiropractic.

The major danger of chiropractic treatment is cervical manipulation, in which the neck is actively moved beyond its physical range of motion by the chiropractor. This can lead to serious damage to the surrounding structures including blood vessels and nerves: patients have developed quadriplegia as a consequence. The blood vessels in the neck (the vertebral arteries) that supply the brain with blood may suffer tearing (dissection) following chiropractic manipulation: the resultant disruption to blood flow is likely to have serious consequences, including permanent brain damage or even death.

A recent systematic review examined adverse events following cervical manipulation (Kranenburg et al. 2017). A total of 227 patient cases were included all of whom had experienced adverse effects. Neck pain was the most frequent indication for which patients had sought treatment. Most (66%) of these patients had been treated by chiropractors. Artery dissection was reported in 57% of the cases. Other complications were disc rupture, spinal cord swelling and blood clots. The most frequently reported symptoms included disturbance of voluntary control of movement, pain, partial paralysis, and visual disturbances.

Another review summarized published cases of injuries in China associated with cervical manipulation (Wang et al. 2012). A total of 156 cases met the inclusion criteria. The problems included soft tissue injuries, dislocated vertebrae in the neck, spinal fractures, nerve damage (including the spinal cord), and syncope (loss of consciousness). These were the (relatively) least severe effects; in addition, particularly serious accidents were reported in 22 patients, including brain hemorrhage, paralysis and, in 5 patients, death.

In 2010, one of us reviewed the deaths reported after chiropractic treatments (Ernst 2010b). This article reported that 26 fatalities had been published in the medical literature. The alleged pathology usually was a vascular accident involving the dissection of a cervical artery. It is important to recognize that many more deaths may have remained unpublished. Clinical trials of chiropractic often omit any mention of adverse effects, thus violating publication ethics and contributing to a misleading representation of chiropractic as a safe modality (Ernst and Posadzki 2012).

As there is no monitoring system, it is impossible to be certain of the extent to which chiropractic damage is underreported. However, a 2001 case-report study by one of us [Ernst] provides some clues (Stevinson et al. 2001). In this work, all members of the Association of British Neurologists were asked to report cases referred to them of neurological complications occurring within 24 h of cervical spine manipulation, over a 12-month period. The response rate was 74%, with 24 respondents reporting at least one case each, contributing to a total of about 35 cases. The reported cases included 9 cases of stroke (4 with confirmation of artery dissection); 3 cases of spinal cord damage; 3 cases of nerve root damage; and 1 case of subdural haematoma (bleeding within the skull). In this study, under-reporting was precisely 100%; in other words, none of the cases discovered in this work had been recorded anywhere else. Given the serious extent of underreporting of chiropractic iatrogenesis, we believe that the true figures for serious injuries and deaths might realistically be brought up by one or even two dimensions.

In their defence, chiropractors often cite the Cassidy study, which gathered data (from Ontario hospitals) on cases of vertebral artery dissection and stroke (Cassidy et al. 2008). Noting that such events *are very rare in the population*", this study reported a positive correlation between [a] consulting a chiropractor or primary care physician and [b] the occurrence of vertebral artery dissection and stroke. However, the authors concluded that this association was explained by patients seeking medical attention for what turned out to be symptoms of a developing dissection (such as headache and neck pain), as opposed to the treatment having caused the stroke. In other words, it was claimed that there is no evidence of a causal association between cervical manipulation and vertebral artery dissection.

That, of course, was what chiropractors longed to hear (and it remains the main basis for their denial of risk). However, a recent paper reviewed the Cassidy et al data and discovered a major problem: there had been a "gross misclassification" of patient cases (Cai et al. 2014). This latter paper re-analysed the data and concluded that, contra Cassidy et al, there is in fact evidence for a causal association between spinal manipulation and arterial dissection, particularly for patients aged below 45 years.

Although the risks of spinal manipulation are routinely downplayed by chiropractors—as evidenced, for instance, by the above quote from the BCA referring to the supposedly 'safe' and 'gentle' nature of chiropractic manipulation—we would hope that at least some chiropractors fulfill their duty to properly explain the dangers to adult patients, so that they can choose whether to accept the risks. However, the use of chiropractic manipulation in young children and babies is ethically much more troubling, because these patients are not competent to assess the risks, and therefore cannot give informed consent.

As we saw above from the conclusion of the Cassidy data re-analysis, the risk of serious injury appears strongest for patients aged under 45. For the very youngest patients, the risks are particularly high: the immature spine of a child or baby is especially vulnerable to damage from manipulation, with the cartilaginous growth plates of pre-pubescent children being particularly at risk of rupture.

Little academic attention has been paid to the risks of using chiropractic manipulation techniques specifically on children. The incidence of such damage is surely under-reported because of the invisible nature of many of the injuries, the effects of which may not manifest until several years following treatment. But the little published research that does exist gives much cause for concern. For example, a 2007 systematic review uncovered 14 significant injuries in children resulting from spinal manipulation (Vohra et al. 2007). Of these, 9 cases were extremely serious (including brain haemorrhage, vertebral dislocation, and paraplegia), and 2 were fatal. Additionally, 20 children were significantly harmed through delayed diagnosis, due to manipulation being used in place of conventional medicine. Seven of these children were impacted by a late diagnosis of cancer, and 2 died due to delayed treatment for meningitis.

As we have discussed previously, there is no credible evidence to support the chiropractic 'subluxation' and 'blocking' theories that underpin spinal manipulation, in children or adults. In light of this fact, and considering both the above risks and the impossibility of informed consent, one might suppose that it would be rare to find chiropractors treating children. Unfortunately, this is not the case. In fact, chiropractic care for children is endorsed by major chiropractic associations, including for example the International Chiropractic Association (ICA) and the American Chiropractic Association (ACA). The reality is that chiropractors commonly treat children for a wide variety of ailments by manipulating the spine, with such treatment frequently beginning soon after a child is born.

This situation is ethically unacceptable. The use of ineffective and dangerous chiropractic techniques on patients who cannot give informed consent should be viewed for what it is, namely fraudulent malpractice (Homola 2016).

In conclusion, while the risks from chiropractic manipulation cannot be quantified, it is clear that serious adverse outcomes, including stroke, quadriplegia and death, can and do occur. Chiropractors assert that these occurrences only affect a small minority of patients; but there can be no doubt that chiropractic harm is chronically underreported. In any case, given that chiropractic is a fundamentally implausible modality, and considering that the evidence of effectiveness of spinal manipulation is highly questionable, all such tragic events are unacceptable.

As with other CAM modalities, a rational explanation of the mechanistic basis of chiropractic manipulation is impossible, thus rendering the notion of informed consent meaningless. The unquantified nature of the risks of chiropractic makes a further mockery of informed consent. From the perspective of medical ethics, the conclusion is inescapable: no consent should mean no treatment. This applies with particular force where children or babies are the chiropractor's intended patients.

5 The Other Options

More and more doctors currently refer patients to CAM practitioners, but referral rates in the other direction remain minuscule. This indicates that CAM therapists rarely guide their patients towards other, more effective therapeutic options. In all fairness, one has to say they rarely can; they are not normally educated to have sufficient knowledge to do so.

A concrete example might explain. Imagine a patient consulting a Reiki healer because she feels tired and suffers from frequent headaches. We would predict that the healer simply administers Reiki and promises symptomatic improvement. The patient's symptoms might indeed get better, perhaps due to a placebo effect. But that would last only a short period of time. For several reasons, the healer cannot consider other therapeutic options:

- He is not trained to diagnose the symptoms of his patients.
- He has no knowledge of treatments other than Reiki.
- He knows nothing about the natural history of his patient's condition.

Let's assume that our patient is experiencing the first symptoms of a brain tumour. The weeks she is wasting receiving Reiki might render her eventual prognosis much worse; in extreme cases, it could cost her life.

This example explains how many CAM practitioners are simply in no good position to inform patients about other, possibly better therapeutic options than their own. In many cases, they cannot even inform the patient what would happen if no treatment was given at all. The natural history of many conditions is such that they get better or disappear completely even without therapy. The best example is probably acute low back pain, one of the most frequently seen conditions by many CAM therapists. We know that it would completely disappear within a week or two in about 80% of patients, no matter what they do and even if they decide to have no treatment at all.

Many CAM practitioners might be aware of this fact, yet they fail to inform their patients about it. The reason is simple: information about other options is not good for business.

6   Conclusions: Informed Consent, Ethics, and CAM

Informed consent is a cornerstone of medical ethics. It is built upon the philosophical concept of autonomy, a principle that is almost universally considered to be one of the foremost pillars of ethics. Informed consent is also supported by utilitarian reasoning, because allowing people the freedom to make their own decisions maximises utility. This is so for several reasons, including: [a] (competent) individuals generally know better than anyone else what is best for themselves; [b] allowing individuals to make their own decisions enables them to learn from their mistakes and thus become better decision-makers (i.e. better maximisers of personal utility); and [c] individuals quite simply feel happier if they know that they are free to choose for themselves. Additionally, the theory of virtue ethics supports the concept of informed consent, because to have any possibility of being a virtuous agent, the individual must be given maximum freedom to make their own decisions.

Some ethicists (albeit a minority) have argued for a restricted concept of informed consent. It has been claimed that patients are best served by a limited form of paternalism, on the basis that healthcare experts are better qualified than the patient to take critical medical decisions. On this view, the patient (despite being competent) is likely to lack the knowledge and insight available to the expert, and may also be distracted by their medical condition. Ethicists who consider autonomy to be an inviolable principle will not be persuaded by this argument. Most utilitarians will also reject it, on the basis that autonomy (at least in the domain of healthcare) maximises utility.

Nevertheless, a utilitarian defence of medical paternalism can be mounted, on the basis that doctors and other healthcare professionals really do know best, to the extent that patients will, in the long run, end up happier if key medical decisions are made for them by the experts. We consider this argument to be unpersuasive, because it implies without justification that patients are not fully autonomous, and rather arrogantly assumes that healthcare professionals are ethically entitled to decide for others.

Nevertheless, let us suppose for the sake of argument that medical paternalism does in fact lead to greater utility, by maximising positive medical outcomes. If this were true, then purely utilitarian ethicists would find it difficult to raise convincing arguments against the notion of a restricted version of informed consent. From this, one can envisage a hypothetical scenario in which it is considered ethically acceptable for doctors to act paternalistically. For example, it might be acceptable to present the nervous patient with an overly positive description of the cost-benefit balance of a proposed therapy, to secure their acquiescence. Under this scenario, the question would then become: ought CAM practitioners also be permitted to act based on a limited view of informed consent?

We believe that the answer is a resounding 'no'. While we can reasonably expect that most paternalistic mainstream doctors are genuine experts in their field, and can be relied upon to make medical decisions based on dispassionate and intelligent analysis of the relevant evidence, the same can hardly be said for most CAM practitioners. Dogma and ideology, not rationality and evidence, are the drivers of CAM practice. To sanction paternalism by CAM practitioners would be to court disaster, as decisions intended 'in the patient's interest' would be made not by impartial experts but by biased and deluded practitioners, with a concomitant lack of understanding on the part of the patient. The resultant suboptimal or catastrophic clinical outcomes would serve to reduce utility, and thus any watering down of informed consent in the domain of CAM could not be supported by even the most paternalistic utilitarian ethicists.

In practice, the reality is that conventional medicine is used in a manner that upholds the principle of autonomy: while inevitably the individual actions of a few doctors and other professionals may be covertly paternalistic, the need for informed consent is held as genuinely sacrosanct in modern medical practice. By contrast, CAM practitioners routinely violate the tenets of informed consent—and this is of course a serious ethical issue.

6.1 The Failure of CAM to Meet the Requirements of Informed Consent

We began this chapter by noting that there are essentially 4 facets of informed consent:

   i. all relevant information must be disclosed to the patient;
  ii. the patient must fully understand what they have been told;
 iii. the patient's decision must be free from coercion or manipulation;
 iv. the patient must be competent, i.e. they must possess decision-making capacity.

We shall conclude by briefly summarising how well CAM practice matches up to these component parts of informed consent.

i. *disclosure of all relevant information*

There is near-universal failure to fulfil this requirement. Most CAM modalities are implausible and lack reliable evidence of effectiveness. Additionally, despite the assumed 'safe and gentle' nature of CAM, some therapies carry significant risks. If practitioners were to be honest and properly explain all this to patients, they would soon be out of business! Of course, some (perhaps most) CAM practitioners may simply be delusional as opposed to dishonest, but the effect is the same in terms of a failure to disclose all relevant information.

ii. *full understanding*

Because most CAM modalities are based upon bizarre, illogical ideas about the causes of disease, and entail treatments that have no scientific basis, CAM practitioners are simply unable to provide patients with valid explanations of the mechanisms by which the treatment will operate, in relation to the underlying cause of the disorder, which must also be accurately explained to the patient. As a result, patients treated with CAM cannot be considered to possess a full understanding of their diagnoses and treatments.

iii. *freedom from coercion or manipulation*

In the context of CAM, this requirement is closely allied to requirement (i) above. CAM practitioners who are fully aware that their modality does not work nevertheless must persuade patients to submit to treatment, using false information to secure acquiescence. This amounts to coercive or manipulative behaviour.

iv. *patient competence*

Perhaps we can at least tick off this last requirement, on the basis that there is no reason to assume that CAM patients lack capacity. While this will usually be the case, it is nonetheless important to recognise that CAM practitioners frequently treat children and non-competent adults (such as those with dementia). In fact, some parents are particularly drawn to CAM treatments for their children, because of the supposedly 'safe and gentle' nature of CAM. In conventional healthcare, 'informed permission' (usually from the parents) can be used as a proxy for informed consent. However, the above very substantial problems serve to undermine informed permission to just the same extent as they undermine informed consent.

6.2 Concluding Remarks

Genuine informed consent is unattainable for most CAM modalities. This presents a serious and intractable ethical problem for CAM practitioners. Attempts to square this circle by watering down or redefining the criteria for informed consent are ethically indefensible. The concept of informed consent and its centrality in medical ethics therefore renders most CAM practice unacceptable. Conventional healthcare subscribes to the ethical principle 'no consent, no treatment': we are not aware of the existence of any good reasons to excuse CAM from this dictum.

References

Allen J, Montalto M, Lovejoy J et al (2011) Detoxification in naturopathic medicine: a survey. J Altern Complement Med 17(12):1175–1180. doi:10.1089/acm.2010.0572

BAC (2017a) Are there any side effects? Top 10 things to know. British Acupuncture Council. https://www.acupuncture.org.uk/public-content/public-faqs/229-are-there-any-unpleasant-side-effects.html. Accessed 26 Apr 2017

BAC (2017b) Research Fact Sheets. British Acupuncture Council. https://www.acupuncture.org.uk/category/a-to-z-of-conditions/a-to-z-of-conditions.html. Accessed 15 Apr 2017

Barrett S (2017) How the "Urine Toxic Metals" test is used to mislead patients. Quackwatch. https://www.quackwatch.com/01QuackeryRelatedTopics/Tests/urine_toxic.html. Accessed 12 May 2017

BCA (2017) About chiropractic. British Chiropractic Association. https://chiropractic-uk.co.uk/chiropractic-care/. Accessed 26 Apr 2017

Bergqvist D (2013) Vascular injuries caused by acupuncture. A systematic review. Int Angiol 32(1):1–8

Bornhöft G, Matthiessen PF (eds) (2011) Homeopathy in healthcare—effectiveness, appropriateness, safety, costs. Springer, Berlin

Bussieres AE, Sales AE, Ramsay T et al (2013) Practice patterns in spine radiograph utilization among doctors of chiropractic enrolled in a provider network offering complementary care in the United States. J Manip Physiol Ther 36(3):127–142. doi:10.1016/j.jmpt.2013.04.002

Cai X, Razmara A, Paulus JK et al (2014) Case misclassification in studies of spinal manipulation and arterial dissection. J Stroke Cerebrovasc Dis 23(8):2031–2035. doi:10.1016/j.jstrokecerebrovasdis.2014.03.007

Cassidy JD, Boyle E, Cote P et al (2008) Risk of vertebrobasilar stroke and chiropractic care: results of a population-based case-control and case-crossover study. Spine (Phila Pa 1976) 33(4 Suppl):S176–83. doi:10.1097/BRS.0b013e3181644600

Cavanagh R (2017) What is the Bowen technique? Rae Bowen therapy. https://www.raebowentherapy.co.uk/bowen-technique/. Accessed 19 May 2017

Charity MJ, Britt HC, Walker BF et al (2016) Who consults chiropractors in Victoria, Australia? Reasons for attending, general health and lifestyle habits of chiropractic patients. Chiropr Man Therap 24(1):28. doi:10.1186/s12998-016-0110-2

Chung KF, Yeung WF, Kwok CW et al (2014) Risk factors associated with adverse events of acupuncture: a prospective study. Acupunct Med 32(6):455–462. doi:10.1136/acupmed-2014-010635

Der-Ohanian T (2008) Homeopathy for cancer. Cancer and homeopathy. http://www.alternativehealth.co.nz/cancer/homeopathy.htm. Accessed 26 Apr 2017

Dougherty PE, Karuza J, Dunn AS et al (2014) Spinal manipulative therapy for chronic lower back pain in older veterans: a prospective, randomized, placebo-controlled trial. Geriatr Orthop Surg Rehabil 5(4):154–164. doi:10.1177/2151458514544956

El-Ghitany EM, Abdel Wahab MM, Abd El-Wahab EW et al (2015) A comprehensive hepatitis C virus risk factors meta-analysis (1989-2013): do they differ in Egypt? Liver Int 35(2):489–501. doi:10.1111/liv.12617

Ernst E (2017) Homeopathic optimism: the case of the 'Swiss report'. Edzard Ernst | MD, PhD, FMedSci, FSB, FRCP, FRCPEd. http://edzardernst.com/2014/05/homeopathic-optimism-the-case-of-the-swiss-report/. Accessed 26 Apr 2017

Ernst E, Pittler MH, Wider B (2006) The desktop guide to complementary and alternative medicine: an evidence-based approach, 2nd edn. Mosby, St Louis, MO

Ernst E (1998) Chiropractors' use of X-rays. Br J Radiol 71(843):249–251. doi:10.1259/bjr.71.843.9616232

Ernst E (1999) Iridology: a systematic review. Forsch Komplementarmed 6(1):7–9. doi:10.1159/000021201

Ernst E (2007) Herbal medicine: buy one, get two free. Postgrad Med J 83(984):615–616. doi:10.1136/pgmj.2007.060780

Ernst E (2009) Chiropractic maintenance treatment, a useful preventative approach? Prev Med 49(2–3):99–100. doi:10.1016/j.ypmed.2009.05.004

Ernst E (2010a) Acupuncture—a treatment to die for? J R Soc Med 103(10):384–385. doi:10.1258/jrsm.2010.100181

Ernst E (2010b) Deaths after chiropractic: a review of published cases. Int J Clin Pract 64(8):1162–1165. doi:10.1111/j.1742-1241.2010.02352.x

Ernst E (2011) Anthroposophy: a risk factor for noncompliance with measles immunization. Pediatr Infect Dis J 30(3):187–189. doi:10.1097/INF.0b013e3182024274

Ernst E, Gilbey A (2010) Chiropractic claims in the English-speaking world. N Z Med J 123(1312):36–44

Ernst E, Hung SK (2011) Great expectations: what do patients using complementary and alternative medicine hope for? Patient 4(2):89–101. doi:10.2165/11586490-000000000-00000

Ernst E, Posadzki P (2012) Reporting of adverse effects in randomised clinical trials of chiropractic manipulations: a systematic review. N Z Med J 125(1353):87–140

Ernst E, Schmidt K (2002) Health risks over the Internet: advice offered by "medical herbalists" to a pregnant woman. Wien Med Wochenschr 152(7–8):190–192

Ernst E, Zhang J (2011) Cardiac tamponade caused by acupuncture: a review of the literature. Int J Cardiol 149(3):287–289. doi:10.1016/j.ijcard.2010.10.016

French SD, Charity MJ, Forsdike K et al (2013) Chiropractic Observation and Analysis Study (COAST): providing an understanding of current chiropractic practice. Med J Aust 199(10):687–691. doi:10.5694/mja12.11851

Guo R, Canter PH, Ernst E (2007) A systematic review of randomised clinical trials of individualised herbal medicine in any indication. Postgrad Med J 83(984):633–637. doi:83/984/633 [pii]

Hampton DA, Kaneko RT, Simeon E et al (2014) Acupuncture-related pneumothorax. Med Acupunct 26(4):241–245. doi:10.1089/acu.2013.1022

Homeopathy Information (2017) Which illnesses can be treated with homeopathy? http://homeopathy-information.com/illnesses-can-treated-homeopathy/?utm_source=dlvr.it&utm_medium=twitter. Accessed 15 Apr 2017

Homola S (2016) Pediatric chiropractic care: the subluxation question and referral risk. Bioethics 30(2):63–68. doi:10.1111/bioe.12225

Kranenburg HA, Schmitt MA, Puentedura EJ et al (2017) Adverse events associated with the use of cervical spine manipulation or mobilization and patient characteristics: a systematic review. Musculoskelet Sci Pract 28:32–38. http://dx.doi.org/10.1016/j.msksp.2017.01.008

Langworthy JM, le Fleming C (2005) Consent or submission? The practice of consent within UK chiropractic. J Manipulative Physiol Ther 28(1):15–24. doi:10.1016/j.jmpt.2004.12.010

Liu Y, Pan J, Jin K et al (2014) Analysis of 30 patients with acupuncture-induced primary inoculation tuberculosis. PLoS ONE 9(6):e100377. doi:10.1371/journal.pone.0100377

MacIntosh A, Ball K (2000) The effects of a short program of detoxification in disease-free individuals. Altern Ther Health Med 6(4):70–76

Maher AR, Hempel S, Apaydin E et al (2016) St. John's Wort for major depressive disorder: a systematic review. Rand Health Q 5(4):12

Mueller A (2016) Unheilpraktiker: Wie Heilpraktiker Mit Unserer Gesundheit Spielen. Riemann Verlag, Munich

Murdoch B, Carr S, Caulfield T (2016) Selling falsehoods? A cross-sectional study of Canadian naturopathy, homeopathy, chiropractic and acupuncture clinic website claims relating to allergy and asthma. BMJ Open 6(12):e014028. doi:10.1136/bmjopen-2016-014028

NHMRC (2017) Homeopathy review. Australian Government National Health and Medical Research Council. https://www.nhmrc.gov.au/health-topics/complementary-medicines/homeopathy-review. Accessed 26 Apr 2017

Onlymyhealth (2017) Homeopathy remedies for AIDS. http://www.onlymyhealth.com/homeopathy-remedies-aids-1298540872. Accessed 28 Apr 2017

Rubinstein SM, van Middelkoop M, Assendelft WJ et al (2011) Spinal manipulative therapy for chronic low-back pain. Cochrane Database Syst Rev (2):CD008112. doi(2):CD008112. doi:10.1002/14651858.CD008112.pub2

Schmidt K, Ernst E (2003a) Are asthma sufferers at risk when consulting chiropractors over the Internet? Respir Med 97(1):104–105

Schmidt K, Ernst E (2003b) Complementary/alternative medicine for diabetes. How good is advice offered on websites? Diabet Med 20(3):248–249. doi:884_3 [pii]

Schmidt K, Ernst E (2003c) MMR vaccination advice over the Internet. Vaccine 21(11–12):1044–1047. doi:10.1046/j.1464-5491.2003.00866_3.x

Schmidt K, Ernst E (2004) Assessing websites on complementary and alternative medicine for cancer. Ann Oncol 15(5):733–742

Schulz V (ed) (2011) Rational phytotherapy: a reference guide for physicians and pharmacists, 5th edn. Springer, Berlin

Shahvisi A (2016) No understanding, no consent: the case against alternative medicine. Bioethics 30(2):69–76. doi:10.1111/bioe.12228

Stevinson C, Honan W, Cooke B et al (2001) Neurological complications of cervical spine manipulation. J R Soc Med 94(3):107–110. doi:10.1177/014107680109400302

Stuber K, Lerede C, Kristmanson K et al (2014) The diagnostic accuracy of the Kemp's test: a systematic review. J Can Chiropr Assoc 58(3):258–267

Vohra S, Johnston BC, Cramer K et al (2007) Adverse events associated with pediatric spinal manipulation: a systematic review. Pediatrics 119(1):E275–E283. doi:10.1542/peds.2006-1392

Wang HH, Zhan HS, Zhang MC et al (2012) Retrospective analysis and prevention strategies for accidents associated with cervical manipulation in China. Zhongguo Gu Shang 25(9):730–736

Watanabe M, Unuma K, Fujii Y et al (2015) An autopsy case of vagus nerve stimulation following acupuncture. Leg Med (Tokyo) 17(2):120–122. doi:10.1016/j.legalmed.2014.11.001

White A, Hayhoe S, Hart A et al (2001) Adverse events following acupuncture: prospective survey of 32 000 consultations with doctors and physiotherapists. BMJ 323(7311):485–486

Chapter 6
Truth

A statement is true when it corresponds with the way the world really is. This is the common meaning of truth that we all know. The opposite of a true statement is a falsehood, an untruth, or a lie. As scientists, we insist on empirical evidence, from observations and experiments, as the basis of scientific truth. In the realm of CAM, very many falsehoods exist: indeed, the vast majority of claims made by CAM proponents are untrue. In this chapter, we shall explain how truth is routinely being neglected and violated in CAM, and discuss the key ethical issues arising from this regrettable state of affairs.

1 Forms of Untruth

Why would CAM practitioners peddle untruths? There are several reasons, ranging from honest error to deliberate fraud. An example from the former end of the spectrum might be the homeopath who, having been 'educated' to believe that homeopathy works, prescribes inert remedies in an honest attempt to alleviate the symptoms of a patient who, unbeknownst to the misinformed practitioner, is suffering from undiagnosed cancer. In this case, the modus operandi is based on naivety and lack of adequate knowledge. At the latter end of the spectrum, an example could be the charlatan who sells a 'therapeutic' anti-cancer agent, such as laetrile, to cancer patients, knowing full well that the therapy is completely ineffective as well as toxic. In this case, immoral pursuit of financial gain (i.e. greed) is the modus operandi.

Considering these examples from the perspective of virtue ethics, the behavior of the charlatan who sells laetrile is judged as indubitably reprehensible, since such behavior entails deliberate deception motivated by greed, coupled with a complete disregard for the patients' interests. By contrast, ethical judgements of the deluded homeopath will tend to be more lenient: there has been no deliberate lying to patients, rather the hapless practitioner has acted in good faith, in accordance with

© Springer International Publishing AG 2018
E. Ernst and K. Smith, *More Harm than Good?*
https://doi.org/10.1007/978-3-319-69941-7_6

the (unsound) training received at college. Yet, the consequences are identical in both scenarios: the patient does not receive life-saving treatment and dies.

As we have discussed previously, it can never be sufficient for healthcare practitioners to merely act in good faith. All healthcare professionals have a positive moral duty to ensure that the treatment of their patients is based upon sound evidence and theory. Thus, regardless of how sincerely an untrue medical belief is held, the practitioner who acts on false beliefs—however honestly and with good intent—is judged negatively according to virtue ethics. (Arguably, the clinician who believes in quackery, and therefore is more convincing that the one who is motivated by greed, is even more dangerous to patients; in other words, conviction renders a quack not less but more harmful.)

For example, consider a physician who believes that AIDS is not caused by the HIV virus and can be cured by vitamin supplements. When he applies these beliefs to the treatment of an AIDS patient, he is behaving in an ethically reprehensible manner; the fact that his delusions are honestly held does not justify his using vitamins in place of antiretroviral therapy, nor does it excuse him morally from the resultant harm to the patient. It was his responsibility to ensure that he properly understood the scientific truth regarding the cause of AIDS and its effective treatment. Because of his failure, this physician is liable to become the subject of justifiable moral opprobrium.

So, from the perspective of virtue ethics, practitioner behavior that is based on falsehoods is always ethically unacceptable, albeit to varying degrees depending on the motivation and honesty (or otherwise) of the individual practitioner.

At this point, we should note that the position of an individual on the above 'motivational spectrum' may change over time, either for better or—we suspect most frequently—for the worse. Suppose a newly qualified homeopath comes to gradually realize, through experience and self-directed inquiry that the foundations of the profession in which she trained are built on sand. What then are her options? There are three main possibilities: [1] abandon homeopathy and retrain for a different profession; [2] try to convince herself that somehow homeopathy does work for her patients, if only through the placebo effect, and cease inquiring critically into the unsettling scientific truth; or [3] keep practicing homeopathy as a means to pay the bills, in full awareness that she is in effect peddling lies to her patients.

The first possibility is probably the most difficult, and therefore the least likely (neither of us knows more than a handful of practitioners who reacted in this way): our homeopath will have invested significant time and resources in her training, and the costs to her of starting afresh may well be intolerable. Yet only this option would be deemed morally praiseworthy by virtue ethicists. The remaining options, namely acting in bad faith or engaging in outright dishonesty, clearly entail behavior that is lacking in virtue, with the latter being due the greatest deprecation.

Condemnation of medical practice based on falsehoods emanates not only from virtue ethicists but also from those who subscribe to different ethical principles. Utilitarians are bound to strongly support the notion that adherence to and pursuit of truth in medicine is ethically mandatory. This is because positive consequences are likely to flow in direct proportion to the extent that adherence to truth is held as a

central principle amongst all agents involved in healthcare, and negative consequences are likely to flow from the converse.

However, utilitarianism yields a slightly different conclusion in judgement of the
above motivational spectrum to that of virtue ethics. In all cases ranging from the
honestly deluded practitioner to the out-and-out charlatan, utilitarian reasoning
delivers a damning verdict: these behaviors are all ethically unacceptable, because
the consequential risks of patients being seriously harmed are high. It follows that a
moral duty exists to act to prevent the patient-harming activity of quack practitioners, regardless of the reasons (honest delusion vs deliberate charlatanism) for
their aberrant practices.

This utilitarian conclusion raises the question: who should act to fulfil the duty of
preventing falsehood-related harm? The answer is that responsibility resides with a
range of agents, operating at different levels. At the highest level, the case can be
made that government ought to create legislation such that medical practice based
on misinformation is outlawed. To some extent this is already done; for example, in
the UK it is illegal to advertise ineffective cancer 'cures'; and most countries have
laws regarding what can and cannot be promoted or sold as medical treatments.
Nevertheless, we consider that, in many or most jurisdictions, there is scope for
such laws to be enforced, strengthened and widened—a theme to which we shall
return in the following chapter.

At an intermediate level, an ethical obligation to act against quackery resides
with various agencies charged with delivering and regulating healthcare and
medical research, such as the NHS in the UK or the NIH in the US. A parallel
obligation resides with colleges and universities involved in educating healthcare
professionals. To maximize utility, these organizations need to enact policies and
procedures that encourage science-based practice and proscribe activities that are
not based on scientific truth.

At the level of clinical practice, an ethical duty resides with individual healthcare
professionals to act when they realize that a colleague is practicing in an ineffective
or dangerous manner. For example, if a surgeon learns that a colleague carries out
inappropriate or unnecessary operations, we would expect the surgeon to take
appropriate action against their wayward colleague. This could take the form of
reporting the individual to the relevant management or professional body.

This ethical duty to report bad practice is a general one, applying to all
healthcare professionals. It follows that doctors, nurses and other medical professionals have an ethical duty to act when they see quackery in their ranks. For
example, if an oncologist were to learn that her colleague uses homeopathy to treat
cancer patients, she would be under an ethical obligation to take the matter further.

The regrettable fact is that, despite the ubiquity of CAM modalities and their
intrusion into conventional healthcare settings, cases where a professional has acted
against a quackery-practicing colleague are extremely rare. This may in part be
explained by the human tendency to place personal or tribal loyalty above broader
ethical considerations such as preventing patient harm. For example, as several
notorious cases attest, rogue surgeons who cause considerable damage to patients

are often not reported by colleagues for years, out of misplaced loyalty and a reluctance to act against one's own ranks.

In cases where quackery (as opposed to plain bad practice) is involved, we suggest that an additional factor may work to further inhibit professionals from performing their ethical duty to act against a quackery-practicing colleague: namely the semi-religious status that is often accorded to many CAM modalities. Because CAM is not science-based, belief in its effectiveness can be almost religious in nature. Where this is the case, there may be a reluctance amongst conventional healthcare professionals to question their colleague's 'faith' in CAM.

While a mainstream professional might take action against a colleague who practices quackery, this almost never happens in the domain of CAM itself. For instance, suppose that an acupuncturist learns that a colleague has been adding some homeopathic techniques when treating patients. In this case, we can be virtually certain that the acupuncturist will not act against his colleague. The main reason for this is simple: a quack is hardly in a position to accuse others of quackery!

Finally, at the lowest but arguably most crucial level, utilitarian reasoning enjoins each professional to self-regulate their own behavior, such that all practice undertaken, and all statements made, are fully compatible with sound scientific reasoning and evidence. At this level, utilitarian ethics and virtue ethics are united in their conclusions.

2   Truthfulness of Some CAM-Researchers

Having dealt with proponents of CAM for many years, we recognise a common theme in their responses to claims of therapeutic ineffectiveness: such claims are met with untruths. In these encounters, we often wonder: are the paraded falsehoods being expressed with honest sincerity, or in a conscious attempt to deceive?

The answer is of little relevance, except to virtue ethicists. To those of us whose ethical outlook is essentially utilitarian, the crucial issue is the harm likely to accrue from the failure to adhere to scientific truth, rather than the motivations of those who promulgate falsehoods.

When a scientist shows evidence to a group of CAM-researchers that their favourite CAM-therapy (for instance, chiropractic) is not as safe and effective as they imagine or pretend, a number of typical responses are elicited. Below are ten such common responses. Each is based on a falsehood or misleading argument, thus rendering the response invalid. (We, the two authors of this book, are well aware that these responses will be used against us as soon as our book is published!)

1. They will state that there is evidence to the contrary.
2. They will suggest that the existing evidence has been misquoted.
3. They will say that medical research is generally so flawed that it cannot be trusted.

4. They will claim that scientific evidence is overruled by centuries of experience.
5. They will reverse the burden of proof.
6. They will say that a new scientific paradigm is required to explain how CAM works.
7. They will claim that scientific evidence and reasoning are not applicable in CAM.
8. They will point out how safe or inexpensive CAM is compared to conventional medicine.
9. They will suggest that the critic is paid by big pharma to defame CAM.
10. They will launch personal attacks on their critics.

2.1 Attempts to Refute Contradictory Evidence

When presented with scientific evidence that their favourite CAM therapy is ineffective, it is common to find the CAM advocate countering this with the claim that evidence exists to the contrary. It is generally easy to dip into the published academic literature and find some papers that seem to 'prove' the effectiveness of just about any therapy. But, as we discussed previously, individual positive reports of clinical effectiveness of CAM therapies are very frequently flawed or irreproducible. Moreover, it is essential to consider the totality of the reliable evidence— and this is, as previously discussed, not favourable for the vast majority of CAM therapies. Most CAM advocates are happy to ignore these inconvenient facts, as they attempt to defend their favoured modality, and there is a plethora of evidence which demonstrates this to be so.

For instance, a survey of 223 professional CAM organisations asked for which conditions they considered their specific therapies to be effective (Long et al. 2001). Responses were received from CAM organizations representing 12 therapies: aromatherapy, Bach flower remedies, Bowen technique, chiropractic, homoeopathy, hypnotherapy, magnet therapy, massage, nutrition, reflexology, Reiki and yoga. The top seven common conditions deemed to benefit from all 12 therapies were, in order of frequency: stress and anxiety; headaches (including migraine); back pain; respiratory problems (including asthma); insomnia; cardiovascular problems; and musculoskeletal problems. For most of these conditions, the therapies in question are not of proven effectiveness. One has to assume that the professional CAM organisations did know this; yet they seemed happy to ignore this fact.

An alternative tactic used by CAM apologists in attempting to refute bona fide and inconvenient evidence is to suggest that such evidence has been misquoted. A common ruse to this end is to cite from original studies one or two sentences which seem to indicate that they are correct. We have found that any reminders that these quotes are out of context usually fall on deaf ears.

2.2   Attempts to Circumvent Inconvenient Evidence

Faced with incontrovertible evidence that their revered modality is unproven, CAM advocates often respond by pointing out that conventional medical research can be less than rigorous, attempting to use this fact to claim that scientific evidence in general is unreliable. This 'tu quoque' ('appeal to hypocrisy') fallacy is popular for distracting from the often embarrassingly negative evidence in CAM—never mind that problems in the aviation industry are no argument for using flying carpets.

A prime example of this tactic is provided by a particularly notorious case involving homeopathy research. At the centre of this story was Jacques Benveniste, senior director of an immunology research facility at the respected French medical research organization INSERM. In the late 1980s, Benveniste submitted a paper to the prestigious scientific journal *Nature,* in which he reported that ultra-dilute solutions of antibodies could activate blood cells known as basophils to produce an immune response (Ball 2004; Davenas et al. 1988).

This was an astounding claim, since it was clear that there should be no anti-bodies present to interact with the basophils, due to the extreme dilution factors employed. Yet the data and methodologies on which Benveniste's paper were based appeared prima facie robust, and the implications of his findings were of course most profound, suggesting that publication in *Nature* was warranted. However, considering the astonishing results, the journal set a condition for publication: a team was to be permitted to undertake a thorough on-site investigation of Benveniste's laboratory and the experimental procedures used to obtain his incredible data.

This investigation revealed multiple weaknesses in Benveniste's experiments, including sloppily executed laboratory work, failure to eliminate systematic errors, and observer bias. Moreover, positive data sets generated during the investigation had implausibly low error rates (statistically, these results were simply 'too good to be true'). Finally, on several occasions, the experiments failed to yield any positive results, for no discernible reasons; in other words, Benveniste's claims had failed one of the most fundamental of all scientific tests: that of reproducibility.

Faced with this damning evidence from the *Nature* team's investigation, one might have expected Benveniste to have admitted that he had gotten it wrong. Alas not, as is clear in the following quote from the subsequent *Nature* write-up of the affair:

> Benveniste acknowledged that his experimental design may not have been "perfect", but insisted (not for the first time) that the quality of his data was no worse than that of many papers published in Nature and other such journals.

> (Maddox et al. 1988)

So, here we observe one of the classic defensive responses of the academic quack: science in general is imperfect, therefore scientific evidence that runs counter to the CAM proponent's expectations is merely on a par with pro-CAM 'evidence'—regardless of the weaknesses of the latter.

This is an argument that, in the final analysis, distils to 'anything goes'. If this is the case, then debates over the effectiveness of CAM are reduced to an interminable lobbing of evidence between scientists and CAM proponents, with the objective truth being unattainable. Of course, such a situation suits the quacks: no matter how much good scientific evidence is amassed to show that a given CAM modality is ineffective, this can always be repudiated by 'alternative evidence' obtained through faulty studies. After all, so the argument goes, because science is always imperfect, no one can say that one set of evidence trumps another.

2.3   Attempts to Evade the Need for Scientific Evidence

The notion that objective truth is unattainable is taken one step further by an academic creed known as postmodernism. Postmodern scholars claim that the objective truth is not merely unreachable, but may not even exist. A detailed consideration of the impact of postmodernism on science is beyond the scope of this book[1]; in passing, we shall simply note that this anti-science doctrine has been embraced enthusiastically by many CAM advocates. The reason for this is clear: if objective truth is a fallacy, then CAM beliefs are immune to scientific challenges.

In reality, while postmodern philosophers may be content with the conclusion that truth is a fiction, CAM proponents do believe in one truth: that their favoured modality actually works. This represents a contradiction for CAM advocates who reject scientific evidence; they want us to believe that their modality is clinically effective, yet they refuse to take seriously the kind of scientific inquiry that could settle the question.

How is this circle squared? The CAM proponents' answer is that practice ought to be guided by subjective *personal experience*. From this perspective, the evidence-base of CAM comprises the testimony of individual patients, who relay changes in their symptoms to practitioners, who thereby gain insights into what 'works'. This amounts to a reliance on anecdotes as the basis for establishing 'truth' in CAM. Previously, when describing the features of good medical research, we criticised those who make general claims based on anecdotal accounts, pointing out the myriad weaknesses of such evidence. In fact, cornerstone medical research tools, such as the RCT, were developed specifically for negating the subjectivity associated with 'personal experience'. This was necessary because such 'evidence' was grossly misleading. Yet many CAM proponents want to blithely abandon scientific methodology and the evidence gained from it, and instead depend upon the personal experiences of individual patients as the means to establish medical truth. In our view, this represents a great and dangerous step back into the dark ages.

---

[1]Readers interested in the reach of postmodernism into science are recommended to start with *Fashionable Nonsense: Postmodern Intellectuals' Abuse of Science* (Sokal and Bricmont 1999), a landmark book based on an exposé of the gullibility of postmodern academics in accepting nonsensical ideas.

2.4   Centuries of Experience Prove Effectiveness

Faced with discomfiting evidence of therapeutic ineffectiveness, many CAM enthusiasts respond with the claim that withstanding the 'test of time' proves the worth of a therapy. We considered this argument in the previous chapter and concluded that, as with astrology, the fact that our ancestors used pre-scientific forms of medicine (such as acupuncture) of itself does not amount to good evidence for the clinical effectiveness of primitive therapies. The fact that many people have apparently experienced therapeutic benefits from a 'time-tested' CAM can be readily explained by a range of factors, including various non-specific effects and the natural history of the condition. And we can draw attention to several practices, such as astrology, clairvoyance, telekinesis, and water divining that, despite having 'stood the test of time', have been shown to be without basis—thus giving the lie to the claim that historical longevity equates with effectiveness. Yet such explanations are generally ignored by CAM apologists, who are happy to deploy the 'centuries of experience' mantra to sidestep contradictory scientific evidence, and thus evade the truth.

2.5   Attempts to Reverse the Burden of Truth

Suppose a particular CAM therapy has been little researched and thus has not been proven to be effective nor ineffective. For instance, currently there are no clinical trials of Shiasu or crystal healing, which makes these treatments not proven to be ineffective. There are many more such modalities, and CAM advocates like to tell us that we must give these therapies the benefit of the doubt. But this has the logic the wrong way around. Science cannot prove a negative; rather, the onus must always reside with the proponent of a therapy to provide evidence for its effectiveness.

Philosopher Bertrand Russell devised the following analogy, now known as Russell's Teapot (or the Celestial Teapot), to refute the idea that the burden of proof lies with the critic to disprove a claim:

> If I were to suggest that between the Earth and Mars there is a china teapot revolving about the sun in an elliptical orbit, nobody would be able to disprove my assertion provided I were careful to add that the teapot is too small to be revealed even by our most powerful telescopes.
>
> But if I were to go on to say that, since my assertion cannot be disproved, it is an intolerable presumption on the part of human reason to doubt it, I should rightly be thought to be talking nonsense.
>
> (Russell 1952)

Russell's Teapot was originally deployed in the specific context of the question 'is there a God?', but it also applies as a general rule. The analogy is highly apposite in the domain of CAM, where anyone is free to make any claims that they wish.

One could, for instance, claim that strokes are prevented by wearing a hessian skull cap, or that drinking boiled jellyfish juice reduces arthritic inflammation, or that placing a small amethyst in one's sock cures influenza. These examples of possible CAM therapies are preposterous—and it would be equally preposterous to claim that these therapies should be presumed effective unless evidence exists to prove their ineffectiveness.

2.6 Claims that a 'Paradigm Shift' Is Needed

The history of science contains many instances where surprising experimental or observational results have led to established theory being overturned. In the 1960s, philosopher of science Thomas Kuhn argued against the then dominant view that science moves ahead by incremental improvements of existing theories, powered by a gradual accumulation of data. Instead, Kuhn claimed that real progress in science occurs in a revolutionary manner, in which a longstanding theory is supplanted with radically a new one, in the face of sustained resistance from the 'old guard' of scientists, who are psychologically wedded to their soon-to-be-ousted theories. Kuhn referred to the adoption of a radically new theory as a 'paradigm shift'—a phrase that has stuck in the academic (and to some extend popular) imagination ever since.

The physics of gravity provides a classic example of a Kuhnian 'scientific revolution': Newton's theory of gravity became dominant in the late 17th century, and worked very well to explain many physical observations, including planetary motion. However, careful observation revealed that certain planetary movements were not perfectly compatible with Newtonian theory; nevertheless, the theory remained dominant for well over 100 years. Then, in the early 20th century, Einstein proposed his theory of general relativity, and this rapidly supplanted Newton's theory. General relativity can explain and predict planetary motion more accurately than Newtonian theory, and for this and several other reasons, is considered to be the superior theory.

Although Kuhn's description of scientific progress has proved highly influential, his depiction has been subject to a number of compelling criticisms. What is perhaps the strongest critique derives from broader observations of scientific progress: while the history of science provides a number of cases that do seem to fit the concept of a paradigm shift (such as the move from Newtonian to Einsteinian theories of gravity), a more realistic picture of science is one in which frequent small revisions to theory are the most frequent drivers of progress.

Another criticism concerns Kuhn's idea that each new theory completely ousts the old. In reality, new scientific theories are more often revised versions of their predecessors. Moreover, scientists who grew up with the old theory will generally be persuaded to change their minds—albeit typically with some resistance and psychological discomfiture—by an accumulation of evidence pointing towards a better theory.

In fact, even in cases where a clear paradigm shift has occurred, it seems that Kuhn's depiction is simplistic. In the case of gravity, the 'supplanted' Newtonian theory is sufficiently predictive for most purposes, and has continued to be used technologically well after Einstein's theory became dominant. For instance, equations derived from Newtonian gravitational theory were sufficiently accurate to guide the Apollo spacecraft to the moon and land on it.

It follows that general caution ought to be used when attempting to apply Kuhnian concepts to instances of scientific progress. Unfortunately, such caution is not always exhibited by the profusion of academics and others who have latched onto the Kuhn's ideas, many of whom apparently see a 'paradigm shift' around every scientific corner. Evidently, many of those who use the term deploy it merely as a buzzword (for example in business meetings), rendering it largely meaningless. Even more concerning is the adoption of Kuhnian semantics by those who assume an anti-science position. This includes certain social scientists and philosophers who assert that scientific theories replace each other not through the accumulation of objective evidence but rather through external influences, such as social pressure (i.e. a successful theory will be one which is most compatible with the dominant political zeitgeist). This is the postmodernist outlook, much beloved of many CAM proponents, which we alluded to above.

The language of Kuhn appeals strongly to CAM proponents when they discover some evidence apparently pointing to the ability of their favoured modality to exert an objectively measurable biological effect, when this would fly in the face of established scientific theory. The rational response to such evidence is to vigorously question it: was there experimental error? Were poor statistical techniques employed to interpret the data? Could fraud have occurred? Have the experiments been independently replicated? By contrast, many CAM proponents adopt Kuhn's clothing and assert that the surprising evidence marks the start of a 'new scientific paradigm', which will be accepted when the scales fall from the eyes of the recalcitrant old guard of scientists.

It is important to recognise the extent to which fundamental scientific theory would have to be revised to accommodate a mechanistic basis for most CAM modalities. For example, considering the extent to which the basic canons of homeopathy run counter to established rules of science and reason, experimental evidence in support of homeopathy would have to be particularly robust. Acceptance of homeopathy would entail (to put it mildly) highly radical revisions of our scientific understanding of the nature of matter itself. Simple arithmetic shows that to receive just one molecule of the diluted agent from a standard homeopathic dilution of 1×10^{30}, the patient would have to consume over 30,000 litres of the homeopathic solution. And many homeopathic medicines are diluted to even greater extremes, ranging up to 1×10^{400}, meaning that to guarantee receiving just one molecule of agent the patient would have to consume more matter than is present within the entire universe. Given that such theory has been assembled through painstaking gathering of evidence involving many scientists and over 200 years of effort, evidence sufficient to demolish our current understanding would have to be exceptionally robust. The notion of a paradigm shift of this magnitude is fanciful.

In the face of astonishing experimental results that conflict with copper-bottomed tenets of science, the only reasonable response is to insist upon particularly high standards of evidence, if the results are to be taken seriously. This central pillar of scientific logic is well expressed in a phrase popularised by astrophysicist Carl Sagan: *extraordinary claims require extraordinary evidence.*

But these logical and scientific considerations seldom appear to trouble CAM proponents. Continuing the example of homeopathy, we return to Benveniste, the researcher who thought he had discovered incontrovertible evidence that ultra-dilute solutions, containing no bioactive molecules, could nevertheless stimulate basophils. A rational response to the experimental data, in line with the concept *extraordinary claims require extraordinary evidence,* would have been for Benveniste to have questioned and reviewed his experimental protocols, laboratory equipment and data analysis procedures. This is because it is most likely that the explanations for the surprising data are to be found in these aspects of the research—as any good scientist knows.

Regrettably and notoriously, Benveniste deviated from established scientific norms and explained his anomalous data by hypothesising that the water used as a diluent could act as a 'template' to retain some kind of imprint of the antibody molecules. This speculation gave birth to the 'memory of water' hypothesis, which we have referred to previously—that fantastical notion offering a supposed mechanistic explanation of how ultra-dilute homeopathic preparations allegedly work.

Anomalous interpretations of spurious data normally disappear, as science progresses and moves ever closer towards the truth. But where CAM is concerned, sadly this is rarely the case. In the case of Benveniste, his research remains revered and celebrated amongst homeopaths and other CAM apologists. Prominent amongst his supporters is Nobel laureate Luc Montagnier,[2] who in 2010 made the following statement during a radio interview:

> He shall have to be rehabilitated, as Jacques Benveniste was right. He was ahead of his time, and scientists find paradigm shifts hard to swallow.

> (Paoli and Freeman 2010)

---

[2]Montagnier was joint winner of the 2008 Nobel Prize in Physiology or Medicine, for the discovery of the HIV virus. Thereafter, he appears to have contracted the 'Nobel disease', in which success apparently goes to the head of the erstwhile respected scientist, resulting in an embarrassing public embrace of crankery. In addition to his views supporting the 'memory of water' concept, Montagnier has embraced 'DNA teleportation', an alleged means by which supposed 'electromagnetic signals' from a DNA sequence can be transmitted to pure water as a 'quantum imprint' such that the water contains the genetic information even though no DNA is present. (The influence of the homeopathic 'memory of water' concept is obvious here.) Apparently, these signals can even be recorded and sent as digital files via email. Prowess in one scientific field does not necessarily qualify one for success in another, and the failure to recognise this reality, coupled with a large dose of overconfidence, may help explain why several other prominent scientists, in addition to Montagnier, after receiving the Nobel prize went on to endorse crackpot ideas, ranging from AIDS denial to alien abduction, and creationism to CAM.

The 'memory of water' concept, usually decorated with various ancillary terms from quantum physics, continues to the present day to be trotted out by homeopaths as the 'scientific' basis of this most ludicrous form of medicine.

We fear that CAM proponents tend to seize upon any research that is favourable to their creed—including results that are incredible—and try to spin them into something of importance, rather than subjecting them to critical evaluation. These ideologues frequently shroud their pro-CAM conclusions in the Kuhnian language of scientific revolutions and paradigm shifts; in doing so they engage in sophistry devoid of intellectual substance. Whether through self-delusion or dishonesty, these CAM advocates are guilty of undermining the scientific truth.

2.7   Claims that Current Scientific Methodology Is not Applicable in CAM

All of the foregoing responses against criticisms of CAM represent various attempts to refute, circumvent, evade or distract from truth that is unpalatable to CAM proponents. However, an alternative strategy can be pursued, entailing rejection of the notion that established scientific methodology, such as the RCT, is of relevance to CAM.

This turn against science is not merely a self-serving reflex by CAM practitioners; despite its innate irrationality, the view that scientific methodology has little or nothing worthwhile to contribute is seriously promulgated in the academic CAM literature.

For example, a much-cited (and therefore influential) 2005 paper makes the following assertions as to how CAM interventions ought, in the opinion of its authors, to be researched (Verhoef et al. 2005):

> The complexity of these interventions and their potential synergistic effect requires innovative evaluative approaches… Classical randomised controlled trials (RCTs) are limited in their ability to address this need… Therefore, we propose a mixed methods approach that includes a range of relevant and holistic outcome measures.

This paper claims that the required approach should be one of 'whole systems research' (WSR), which:

> must not focus only on the "active" ingredients of a system. An emerging WSR framework must be non-hierarchical, cyclical, flexible and adaptive, as knowledge creation is continuous, evolutionary and necessitates a continuous interplay between research methods and "phases" of knowledge.

We confess that we are not quite sure what these words mean. A cynic might conclude that they amount to mere pseudo-profundity, designed to give the illusionary impression that the paper's authors possess great wisdom, while disguising a dearth of meaningful content in their assertions.

Whatever the authors' words are intended to convey, we would draw one central conclusion: because high quality science-based research almost invariably finds no good evidence for the effectiveness of CAM, the paper's authors (and many others of their ilk) want to downgrade or abandon science as a means of establishing truth, replacing it with something altogether less rigorous, and much more 'CAM-friendly'—because this is the only way that CAM therapies can be shown to be 'effective'.

When this anti-science ideology is challenged, some CAM proponents accuse their critics of 'scientism'. This term, as a pejorative, refers to the contention that the only meaningful knowledge claims are those that are based on scientific evidence. For example, one academic defender of CAM writes:

> Taken to the extreme, scientism defaults to Internet-fueled inquisitorial intolerance which, supported by certain academics, sections of the media, and (usually anonymous) blog sites, systematically vilifies anything considered 'unscientific', e.g. the campaign to undemocratically rid Britain's NHS of its homeopathy/CAM facilities.

> (Milgrom and Chatfield 2012)

It is clear to us, and we would hope to most scientists, that looking to science for answers to questions beyond its scope is a mistake. This would apply to several fields of inquiry, for example history, language, and musicology, to name but three. However, we suggest that the application of scientific methodology to medical questions around therapeutic efficacy and safety is entirely warranted, and thus does not constitute 'scientism'. Questions of medical effectiveness—for all treatments, whether mainstream or 'alternative'—properly reside in the domain of science. The onus is on with opponents of this position to provide convincing arguments for the exclusion of science from medicine. To date, we have heard no such arguments.

In fact, we find a great deal of hypocrisy amongst CAM proponents concerning their positions on the usefulness or otherwise of scientific evidence. For example, while many quacks boldly reject the RCT—despite it generally being considered as the 'gold standard' of clinical research—we have yet to find the CAM enthusiast who declines to pounce upon and trumpet any instances where scientific research appears to show some CAM therapy to be effective.

Ironically perhaps, the above authors' denunciation of scientism serves as an example. In the same piece, we find the authors explicitly defending homeopathy on the (supposed) grounds that "compelling scientific evidence" shows it to be effective—thereby (apparently without realising the contradiction) acknowledging the central importance of science to medicine.

A further reason given by CAM researchers for the rejection of scientific methodology is the allegedly 'reductionist' nature of scientific research. We considered reductionism previously, in the context of claims that the allegedly more holistic nature of CAM renders it superior to science-based medicine. In the context of research, reductionism simply means trying to understand phenomena (such as drug effects on the body) by focusing on small levels or components (such as individual cells isolated in the laboratory), abstracted from the larger whole. Opponents of scientific medicine invariably use the term reductionist as a pejorative; yet approaches based on studying small biological components is highly valuable as a means towards medical truth. Ultimately, we want to know how a pharmaceutical drug, herbal concoction or homeopathic remedy interacts with the body's cells, and even with smaller components, such as organelles, molecules and gene sequences. This is because the actual effects of drug molecules occur at the cellular and molecular scales—meaning that the deepest understanding of a drug's precise activity will come from experimentation involving these tiny scales.

Moreover, it is essential that experiments function at an abstract, unnatural level. The whole point of scientific experiments is to be able to isolate (in the laboratory for example) a defined component (such as a cell sample) and alter only one variable at a time, to painstakingly build up reliable knowledge. The much-maligned RCT is also such an experiment where one factor (the specific effect of the tested therapy) is being isolated from a myriad of confounders. To denigrate this approach, as CAM researchers usually do by misusing the word 'reductionist' as an insult, is to misunderstand, or unjustifiably reject, the sort of scientific methodology that has contributed to a veritable explosion of knowledge—and concomitant improvement in human welfare—that has been underway since the Enlightenment.

Moreover, to conduct such 'reductionist' research in no way implies a rejection of the need for higher level investigations. All medical scientists are aware that in order to understand the operation and effectiveness (or otherwise) of any drug, laboratory research using isolated cells and genes must be augmented by research focused on larger components (such as the responses of organs and physiological systems), and ultimately clinical trials and post-marketing surveillance studies. Thus, accusations of 'reductionist' from CAM proponents against scientific methods are sloppy, fall well short of their intended target, and essentially reveal their limited understanding of science.

Where CAM proponents want to downgrade or disregard the role of scientific evidence, the question must be asked: what would they replace it with? Search for an answer in the academic CAM literature, and one finds descriptions of alternative means to the 'truth' that may reasonably be described as gobbledegook. One example is the 'WSR' quote above. Here are a few other typical examples.

In the context of Ayurvedic medicine:

> Relevant research design issues will need to address clinical tailoring strategies and provide mechanisms for mapping patterns of change that account for the contiguous, self-replicating, cumulative, and synergistic theories associated with successful Ayurvedic treatment approaches.
>
> (Rioux 2012)

As with the WSR example, we struggle to understand what of substance—if anything—this statement conveys. It may amaze the reader that this kind of obscurantist language gets past journal editors and peer reviewers; but this can be explained by the self-referential, biased nature of the CAM academic world as discussed previously.

The following paper invents a metaphoric 'wheel of knowledge' as a proposed means of investigating 'healing relationships' in CAM:

> Applying this wheel to the issue of assessing impact in healing relationships reveals the need for multiple methods, perspectives, and triangulations. A critical multiplist strategy is one means for advancing this area of research. A double-helix trial design is introduced, in which one strand consists of a standard quantitative approach and the other consists of qualitative methods. The 2 strands are bonded by the questions addressed and by the participants in the study.

> (Miller et al. 2003)

While the inventiveness of this paper's authors cannot be doubted, and their allusion to the DNA double helix is a nice poetic touch, there is little of substance in the argument. The only feature worth noting is the 'multiplist strategy', to which we shall return shortly.

The authors of our next example want CAM research to go beyond merely looking at clinical 'outcomes':

> The use of complex conceptual models, such as those based on programme theory, are a basis for understanding the complexity of the experience of health-care interventions within the wider social and cultural context of peoples' lives. They are an alternative basis on which to understand and evaluate the changes in health and wellbeing associated with many complementary and alternative medicine interventions and to understand the role of various factors in promoting positive changes.

> These models lead to a different interpretation of 'outcomes' which encompasses the interactions and learning that constitutes the treatment experience over time, and views the patient as an active agent who will interact with an intervention in ways that produce individualised changes.

> (Paterson et al. 2009)

Indeed, why stick to boring old-fashioned RCTs with their troublesome negative outcomes? An 'alternative basis' for understanding therapeutic effectiveness sounds much more reassuring—to the CAM enthusiast at least.

In the following final example, the authors would like to see something called 'State-Space Grid (SSG) Analysis' used as the basis for researching 'whole systems of complementary and alternative medicine' (WS-CAM).

> The SSG method generates a two-dimensional visualization and quantification of the inter-relationships between variables on a moment-to-moment basis.

> Practice theories of WS-CAM encompass the holistic health concept of whole-person outcomes, including nonlinear pathways to complex, multidimensional changes.

> Understanding how the patient as a living system arrives at these outcomes requires studying the process of healing, e.g., sudden abrupt worsening and/or improvements, 'healing crises', and 'unstuckness', from which the multiple inter-personal and intra-personal outcomes emerge.

> (Howerter et al. 2012)

This paper informs us that SSG is, apparently, a *'mathematically less invasive methodology'* than traditional research approaches. While we are pleased that CAM researchers have found a research approach that does not challenge their mathematical inadequacies, we are perplexed and astonished that this feature is viewed as a justification for a particular research methodology by the authors.

As scientists, much of the discourse present in the above papers strikes us as mere 'bullshit bingo', featuring overuse of a host of hollow terms, such as healing, holistic, multidimensional, nonlinear, synergistic, and whole systems. We can't say whether the authors of these papers understand or truly believe what they write. We suspect that there is, in general, a good deal of intellectual dishonesty in the context of 'alternative research' proposals. We say this in view of the convoluted and often pompous language which is almost universally employed in such writing, which seems to serve merely to disguise extremely vague or vacuous concepts.

If one seriously attempts to discern what of substance is actually being suggested by such authors, two basic proposals can be distilled from their pseudo-profound ramblings:

[1] Scientific research should be replaced by *personal experience*.

[2] A broad *mixture* of research methods and forms of evidence should be used.[3]

We considered the use of personal experience above, in the context of attempts by CAM proponents to evade the need for scientific evidence altogether, and we rejected this approach, pointing out that it is tantamount to going back into the dark ages.

By contrast, the call by CAM researchers for *mixed methods* (the so-called 'multiplist strategy') may seem at first sight to be less extreme than a wholescale abandonment of established scientific research methods. But, on closer inspection, this is not so, for several reasons.

Firstly, the CAM proponents' proposed mixture of methods always includes one hopeless category, namely personal experience (i.e. #1 above). Adding a defective component to a set of proven methodologies cannot enhance the likelihood of obtaining reliable knowledge; rather, its inclusion can only contribute misleading information.

Secondly, adding together all the evidence from a mixture of weak and strong methodologies is always going to dilute the more valid evidence with unreliable evidence; this will inevitably yield an overly optimistic picture of CAM effectiveness.

Finally, the implication of a mixed methods approach is that, if one method (such as a rigorous clinical trial) does not give sufficient evidence of effectiveness of a CAM therapy, then credence should be given to other weaker sources of evidence— i.e. those from personal experience. This final recourse is in effect no different from #1 above, i.e. a wholescale rejection of science as the basis for truth in CAM.

---

[3]Proposals for 'mixed method' CAM research are often accompanied with the justification that the use of several methods is non-reductionist in nature; i.e. several levels of phenomena are involved. However, as discussed above, the reality is that scientific medicine already depends on a mixture of rigorous methods (including laboratory experiments and RTCs) in order to establish the effectiveness of a therapy. Thus, the charge of 'reductionism' (as a pejorative) simply does not stick.

2.8 Claims that CAM Is Very Safe Compared to Conventional Medicine

Unable or unwilling to try to refute the truth that there is a dearth of evidence of CAM effectiveness, it is common for CAM proponents to try to shift focus and instead promote CAM on the basis of its presumed safety. This works very well for distracting from the lack of evidence for CAM and regularly convinces lay people.

Previously we have given examples of patients being seriously harmed by CAM treatments, including cases of brain damage, collapsed lung, gangrene, heart failure, quadriplegia, and death. These cases came from published academic research: undoubtedly many more cases of harm have occurred that have not been reported, written up and published. In truth, therefore, it is simply false to claim that CAM is totally 'safe'.

Faced with the reality that CAM can and does inflict harm, its apologists frequently modify their argument into a relativistic form: they claim that, although not free from risks, CAM is much safer than conventional medicine. This is another form of the 'tu quoque' fallacy. It fails, because it ignores the crucial flip side of risk—the benefits. To get to the truth about the value of a therapy, it is always necessary to weight its risks and benefits. While many conventional therapies are undoubtedly hazardous—coronary bypass surgery or chemotherapy for instance—they also provide proven life-saving benefits. By contrast, the benefits of CAM tend to be vanishingly small or non-existent: accordingly, risks to patients weigh much more heavily against CAM, even in the case of a therapy where the degree of risk is demonstrably lower than that of its conventional counterpart.

In the case of some CAM modalities, it is true (at least in principle) that the therapies offered are free from any risk of directly inflicting harm on the patient. Homeopathy is a case in point: homeopathic medicines typically contain no active ingredients whatsoever, and therefore cannot have negative effects in the patient's body.[4] However, to conclude that homeopathy is therefore inherently 'safe' would be wrong: 'treating' a patient who is suffering from a serious condition with a dud medicine is hardly a safe course of action when an effective (conventional) treatment could have been used instead.

---

[4]While in principle ultra-diluted homeopathic medicines cannot cause biological effects in patients, in practice this is dependent upon the product being free from contaminants. In most jurisdictions, CAM medicines are not subject to the strict regulations on manufacturing that apply to conventional pharmaceutical medicines. In the case of homeopathic medicines, the diluent (water or alcohol) could be impure, or if in tablet form the base material could be contaminated. CAM preparations on the market are not routinely tested for contaminants, and only a limited amount of research has been conducted into this issue; but the available results show that it would be dangerous to assume that CAM products such as homeopathic remedies are free from toxic contaminants. For example, one investigation discovered contamination levels above recommended daily maxima for aluminium, arsenic, cadmium, mercury and lead in various CAM medicines (Genuis et al. 2012).

It is possible to discern hypothetical scenarios where CAM is completely safe: for example, a patient with a nuisance condition such as a mild common cold (for which no curative conventional treatment exists) taking an uncontaminated and highly diluted homeopathic preparation. But the safe nature of this type of scenario should not be permitted to serve the aim of the CAM apologist; namely to distract from the lack of evidence of effectiveness of CAM and thus avoid the truth.

2.9   Personal Attacks on Critics

As critics of CAM, we have observed that, whenever seemingly reasonable arguments have been exhausted, the typical CAM proponent is likely to resort to some form of personal attack. There are several such ad hominem arguments available to apologists of quackery, ranging from claims that the critic is unqualified, to allegations of corruption.

Most critics of CAM have not undergone any specific training in CAM therapeutic practice. Thus, it is usually open to CAM apologists to make the easy assertion that their opponent is not competent to criticise. This is a form of 'argumentum ad verecundiam', or argument from false authority. This fallacy has an intriguingly circular and self-serving structure: experienced CAM-researchers are almost invariably believers in CAM; thus, only those who believe in CAM can judge it. The corollary is that any and all criticism of CAM must be invalid! This kind of closed thinking is characteristic of religious cults, and is inimical to genuine efforts to seek the truth.

If the hapless CAM defender fails to make convincing arguments that their opponent ought to be ignored on grounds of competence, then blatant lies (or outright delusions) are likely to come into play: it will be alleged that the critic is corrupt. In discussions about the evidence for CAM, enthusiasts often come up with a simple theory: the research has been done and it has produced fabulous results, but it is being actively suppressed by the enemies of CAM, usually characterised as 'big pharma' or 'the medical establishment'.

According to this theory, the pharmaceutical industry, together with the vast assemblage of scientists and physicians dependent upon it, feel so threatened by the findings of CAM research that they conspire to make it disappear. You see, they have no choice, really; the CAM therapy in question is so effective that it would put big pharma and conventional medicine out of business, if it had not been sabotaged by malevolent scientists.

Is there any truth in these extraordinary claims? When we ask CAM enthusiasts to provide supporting evidence, all that they can give are various instances of malpractice in pharmaceutical research. Of course, cases of serious research misconduct have certainly occurred in the drugs industry, some instances of which have been popularised and are fairly well known.[5] However, it is ridiculous and paranoid to imagine that these notorious individual misdeeds, however severe, amount to a grand conspiracy against CAM.

Similarly, any charge of corruption made against CAM critics requires specific and strong evidence. It would be illogical—and unjust—to assume that those who choose to criticise CAM are ipso facto in the pocket of big pharma. If convincing evidence against the critic is not forthcoming—as is virtually always the case—then the charge fails utterly. However, the CAM defender who makes such an unfounded assertion may nevertheless be successful in undermining the critic—at least in the eyes of gullible CAM enthusiasts, or credulous laypeople—by creating a suggestion of 'guilt by association'.

Ad hominem attacks on CAM critics are commonplace, with allegations ranging from professional incompetence to corruption. These are easy claims to make—but in almost all cases they are devoid of truth. From an ethical perspective, the defence mechanism of airing unsubstantiated accusations against critics is wholly unacceptable.

3   Truth About Some Fundamental Assumptions in CAM

Why do so many people opt for CAM? What is its attraction? Enthusiasts claim, of course, that their favoured CAM modality is popular because it is effective and safe. As there usually is little data to support this claim, it is probably not the true answer. There must be other reasons. For instance, it could be due to CAM proponents' assumptions being false and consumers falling victim to falsehoods, lies and 'tricks of the trade' which CAM practitioners—physicians as well as lay practitioners—use in order to convince the often all-too-gullible public of their offerings.

We have previously considered a number of these falsehoods, including the unscientific notion that the body needs 'detoxed', the fallacious claim that a therapy must be effective if it has 'stood the test of time', the false assertion that CAM treats the 'root cause' of disease, and the fallacy that 'natural is good'. In the following sections, we list a selection of additional falsehoods that are frequently promulgated by CAM practitioners in their efforts to persuade patients to sign up for treatment.

---

[5]For a trenchant account of all that is allegedly wrong with the drugs industry, see the book 'Bad Pharma' (Goldacre 2013).

3.1   Treating a Non-existing Condition

Some CAM practitioners have made a true cult of treating conditions that the patient in question does not have. For example, a patient attending a chiropractor is highly likely to receive a diagnosis of 'subluxation'. The same patient attending a TCM practitioner will most likely be told that there is something amiss with their Qi. They can't all both right! In fact, in both cases they are wrong: neither subluxations nor Qi exist, therefore the diagnoses are figments of the imagination.

Each branch of CAM seems to have created its very own diagnoses. To arrive at such diagnoses, the practitioner will, as previously discussed, often use diagnostic techniques which have either been found to lack validity, or which have never been validated at all. The result can only be a false diagnosis. Either the practitioner has misdiagnosed a real medical condition, or has diagnosed a non-existing condition.

Many perfectly healthy people are prone to symptoms, real or imagined, that trouble them and may result in a visit to a doctor or CAM therapist. Common examples of bothersome symptoms often with no underlying pathology include tiredness, headaches, gastrointestinal discomfort, and back pain. In many cases, these symptoms will disappear of their own accord. For less fortunate individuals, the symptoms will persist or recur; some of these people become the 'heartsink' patients who frequently turn up at their doctor's office with difficult-to-treat symptoms that have no evident underlying cause.

To the CAM practitioner, such patients are easy prey. They can and almost invariably will be given a false diagnosis, the obvious advantage of which is that the practitioner can treat the patient's symptoms again and again—until the client has run out of money or patience. Eventually, the practitioner may announce to the patient "you are now healthy". This happens to be true, of course, because the patient has been healthy all along.

CAM practitioners who give false diagnoses for non-existing conditions are either misinformed to the point of delusion, or have deliberately chosen to deceive their patients. In neither case can they be said to be behaving in an ethical manner. All practitioners have a personal ethical responsibility to ensure that their practice is based on concepts that are truth-based, and of course deceiving patients is morally reprehensible.

3.2   Maintenance Treatment

The term 'maintenance treatment', as used by some CAM practitioners, describes the regular treatment of an individual who is entirely healthy but who, according to the practitioner, needs regular treatments in order to remain in good health. Many chiropractors, for example, proclaim that maintenance treatment is necessary for keeping a person's spine aligned—and only a well-serviced spine will keep all of our body's systems working perfectly. It is like with a car, they claim: if you don't service it regularly, it will sooner or later break down. And for good measure, these chiropractors recommend that patients bring along their families—including infants—for regular maintenance treatment.

For many consumers, this sounds convincing enough to make them fall for it. However, as we set out previously, the value of chiropractic maintenance treatment is unproven and potentially harmful. Once again, a CAM practitioner who offers such 'maintenance treatment' is either not well-informed or is deliberately deceiving patients; and both would, of course, be unethical.

3.3   Stimulating the Immune System

"Your immune system needs stimulating!" is something we hear regularly in CAM. By contrast, conventional clinicians are more reserved about such an aim. In modern medicine, deliberate stimulation of the immune system is a goal only in certain very rare circumstances (for example, in highly specialised procedures aimed at stimulating the immune system to destroy tumour cells). Much more frequently, the opposite effect is required, and physicians will deploy powerful drugs to suppress the immune system (mainly in autoimmune conditions such as rheumatoid arthritis or Crohn's disease, or following organ transplantation).

Responsible clinicians would never want to stimulate the immune system of patients unless a very specific reason existed for doing so. Most patients have no problem with their immune system, and stimulating a healthy immune system is undesirable—and hardly possible in any case. Where responsible clinicians do need to use immune stimulating agents, they would never use any CAM treatment, because alternative 'immune stimulants' simply do nothing whatsoever to increase the activity of any part of the immune system.

Yet practitioners from across all CAM modalities claim that health can be improved or protected by the catch-all stratagem of immune system stimulation. This betrays either a profound lack of understanding of immunology on the part of the practitioner, or a deliberate attempt to blind the patient with (pseudo)science, doubtless in pursuit of income. Either way, CAM practitioners who push (illusory) immune stimulation on their patients are culpable of ethically reprehensible practice.

4   The Problem Is Due to the Poisons Your Doctor Gave You

Some CAM practitioners are eager to advise patients that their symptoms are caused by the poisonous drugs prescribed by their doctor—who is, of course, is in cahoots with 'big pharma'. Many CAM advocates thrive on conspiracy theories, and the evil 'medical mafia' is one of their all-time favourites (Oliver and Wood 2014). It enables foolish or unscrupulous CAM practitioners to instil fear into the minds of their patients, thus minimising the risk of them returning to real medicine. Furthermore, this ploy destroys trust in conventional medicine—which will affect public health, if sufficiently widespread.

4.1 Symptoms Must Get Worse Before the Patient Gets Better

Many patients experience a worsening of their complaints after receiving CAM treatments. To this, CAM practitioners often respond that such an experience is normal or even desirable because things have to get worse before they get better. For instance, homeopaths would expect that the presenting symptom of many patients get considerably worse after the optimal homeopathic remedy has been found and administered (Grabia and Ernst 2003). CAM practitioners tend to call this an 'aggravation' or 'healing crisis'. But this is a phenomenon for which no or very little compelling evidence exists.

Imagine a patient with moderately severe IBS symptoms consulting a CAM practitioner and receiving treatment. There are only three things that can happen to her:

- she can get better,
- she might experience no change at all,
- or she might get worse.

In the first scenario, the practitioner would claim that his CAM therapy was responsible for the improvement. In the second scenario, he might say that, without his therapy, the patient's symptoms would have deteriorated. In the third scenario, he would tell her that the healing crisis is the reason for her experience and elaborate that the healing crisis is essentially a good sign because it signals that the optimal treatment has been administered.

Yet, the so-called healing crisis is a fantasy; it does simply not exist. Presumably some CAM therapists have convinced themselves to the contrary; for others, it is nothing more than a 'trick of the trade' to make money. Again, none of this is ethically acceptable.

4.2 A Cure Takes a Long Time

Imagine a scenario where, even after numerous therapy sessions, a patient's condition has not changed. Let's assume the problem is chronic back pain, and that it has not improved despite the many treatments and the considerable amounts of money spent on it. In such a situation, most patients would discontinue the therapy in question. And this is, of course, a threat to the CAM practitioner's cash flow.

Averting the risk is simple: the practitioner merely needs to explain that a cure cannot possibly be expected to be rapid, in view of the fact that the patient's condition has been going on for quite a long time.

This plea to carry on with the ineffective treatments despite the absence of any improvement of symptoms is usually not justifiable on medical grounds, and one would hope that all but the most deluded of practitioners will call time when it eventually becomes clear that their treatment is not working. Sadly, CAM practitioners are financially incentivized to retain their patients even in the face of demonstrable failure, and there can be no doubt that this is a factor behind the reality that many CAM practitioners keep treating patients who fail to get better. This is yet another instance of inconvenient truth (i.e. CAM ineffectiveness) being sacrificed to suit the CAM practitioner—despite the obviously unethical nature of such behaviour.

4.3   Thinking Holistically

The notion that CAM takes care of the whole person has been mentioned previously; it can be used as a most attractive and powerful ploy. Never mind that nothing could be further from being holistic than, for instance, diagnosing conditions by looking only at a patient's iris (iridology), or focussing on her spine (chiropractic, osteopathy), or massaging the soles of her feet (reflexology). And never mind that any type of good conventional medicine is by definition holistic. To CAM practitioners, the label 'holistic' is a most desirable one, because nothing sells quackery better than holism. Therefore, most CAM practitioners rub holism into the minds of their patients whenever they can, claiming it as unique to CAM.

Holism has the added advantage of supplying seemingly plausible excuses for therapeutic failure. Imagine a patient consulting a practitioner with depression and, after prolonged treatment, her condition is unchanged. To such a situation, the holistic practitioner might respond that he never treats diagnostic labels but always the whole person. The patient's depression might not have changed, but surely other problems have lessened. If the patient then introspects a little, she might find that her appetite has improved, that her indigestion is better, or that her tennis elbow is less painful; some symptoms always change given enough time. Holism may be a lie, but its benefits for the unethical CAM practitioner are obvious.

4.4   Claims that the Placebo Effect Justifies CAM

When their attempts to avoid the truth are all refuted, CAM proponents tend to use one final defence: the placebo effect. Here they are on firm ground, as the placebo effect certainly exists: if almost any form of attention is paid to patients, their self-reported symptoms will improve. Indeed, as we have stressed previously, the double-blind nature of good clinical trials came about as a defence against the placebo effect being mistaken for specific effects of the therapy under test.

Thanks to the placebo effect, almost *any* conceivable therapy will have an effect on at least some of the patients treated.[6] But CAM proponents are reluctant to use the placebo effect as a defence. They may use it as a last resort—when they can no longer deny the truth. Alternatively, they will add it into claims of actual effectiveness, either for good measure, or as a form of insurance against the truth emerging.

CAM apologists are coy about deploying the placebo defence because to do so implies defeat: it would be an admission that their favourite CAM actually does not intercede in the body to bring about biological improvements, but simply 'works' through a common non-specific psychological process. This implicit admission is unpalatable to CAM enthusiasts for several reasons. Firstly, it reduces the CAM therapy to doing merely what *all* medical treatments—including conventional therapies—do anyway (albeit for conventional therapies usually as a minor part of the overall effectiveness package). Secondly, as discussed in Chapter 1, it is well known that the placebo effect tends to be unreliable, short-lived and limited in magnitude—thus, claiming that a therapy works (only) as a placebo is in fact an admission of low effectiveness. Thirdly, only the most certifiably deluded can imagine that the placebo effect is able to bring about any significant improvements to the many fundamental pathologies that afflict humankind (and which myriad CAM therapies purport to treat), such as metastatic cancer, neurodegenerative conditions, or infectious disease. And finally, if CAM therapies work only on the basis of the placebo effect, there remains nothing to distinguish these therapies from each other—or from conventional therapies. Why would a patient choose (say) acupuncture, with all those uncomfortable needles, when they could simply choose one of many other alternative CAM therapies—or a conventional therapy? This threatens the CAM practitioner with a loss of marketable identity.

Nevertheless, in reality it is commonplace for CAM proponents to resort to the placebo defence, as their 'get out of jail free' card. They might say that, even if their therapy is a placebo, it does nevertheless help their patients; therefore, the treatment in question is a good one. For practitioners who do so, there is a central ethical problem: reliance on the placebo effect implies lying to patients. Patients expect their prescribed therapies to have actual effects. A significant placebo effect depends upon such beliefs. An argument could be made in favour of hoodwinking patients in order to facilitate a placebo response; however, such an argument would be highly paternalistic, and would thus conflict with the ethical imperative to facilitate patient autonomy. Moreover, the notion of 'informed consent'—which is, as

---

[6]This does depend on the patient being aware that they have been 'treated', and for some therapies this is not necessarily the case. Intercessory prayer at a distance (covered previously), if used without the patients knowing they are being prayed for, is one such example.

discussed previously, an ethical cornerstone of modern healthcare—demands the positive provision of full information to the patient. Thus, causing patients to believe a falsehood (that a CAM therapy can directly alter physiological functioning) amounts to a denial of informed consent, and is thus unethical. It is also, for these reasons, prohibited in most jurisdictions.

It has recently been argued that 'open label' placebo interventions—where the patient is told that the proposed treatment is a placebo—are ethically acceptable, with no breach of autonomy because the 'deception' has been authorised by the patient (Blease et al. 2016). We have no problem with the internal logic of this argument. However, there is a problem with its implied premise: namely that a placebo effect can be elicited despite the patient knowing full well that their treatment can have no specific effects. The trouble is that this premise is based upon rather minimal and contested evidence. Research in the field of open-label placebos is nascent, and very few clinical investigations into this phenomenon have been published.

The above authors focused on one study: an open-label placebo study for irritable bowel syndrome, this being the first study of its kind (Kaptchuk et al. 2010). Notably, CAM advocates have seized upon this single piece of research as a supposed justification for otherwise ineffective CAM therapies.

In this study, 80 patients were randomised into two groups, a control group and a treatment group. Practitioners told the treatment group patients that they would be treated with:

> placebo pills made of an inert substance, like sugar pills, that have been shown in clinical studies to produce significant improvement in IBS symptoms through mind-body self-healing processes.

The treatment group patients duly received the placebo pills, whereas the control group patients were not treated. The study lasted for three weeks, and the treatment group showed a significantly greater reduction in symptoms compared with the control group.

There are a number of issues with this study, but our main concern is that the patients probably did not fully realise that they were being treated with an entirely dummy agent. We would be surprised if the average patient truly understood the term "inert" in the context above, nor the phrase "mind-body self-healing processes".

So, while this study is of interest, it seems clear that future research of this kind will have to be better designed (and replicated) before any confidence can be placed in the use of open-label placebos. Accordingly, we consider that any move to alter ethical guidelines to permit the use of open-placebo treatments in medical practice would be premature.

We conclude that, as per existing codes of medical ethics, it is usually ethically unacceptable to treat patients with placebo-only treatments, as it involves deception and a concomitant violation of the principle of patient autonomy. CAM practitioners who resort to doing so are in breach of a central tenet of medical ethics; furthermore, they are displaying intellectual bad faith and demonstrating that they have little regard for truth.

5   High Profile Purveyors of CAM Falsehoods

In the realm of CAM, consumers are regularly—and we would argue systematically—misled by falsehoods. This abundance of misinformation is one of the main reasons why CAM is currently popular. The 'alternative facts' are currently being disseminated on virtually all levels: CAM practitioners are obviously at the forefront but charities, manufacturers of CAM products, researchers and government bodies are contributing regularly. Even royalty can be a contributor.

The most prominent advocate of CAM in the UK is H.R.H. Prince Charles (the Prince of Wales). He regularly makes public pronouncements supporting the use of CAM on the NHS, and even wrote to the Health Secretary stating he

> couldn't bear to see people suffering unnecessarily when a complementary approach could make the difference.
>
> (Robinson et al. 2015)

Charles' support for CAM therapies is well-documented: in 2006, for instance, he was invited by the WHO to elaborate on his most bizarre concepts in relation to 'integrated medicine'. He told the World Health Assembly:

> The proper mix of proven complementary, traditional and modern remedies, which emphasises the active participation of the patient, can help to create a powerful healing force in the world...Many of today's complementary therapies are rooted in ancient traditions that intuitively understood the need to maintain balance and harmony with our minds, bodies and the natural world...Much of this knowledge, often based on oral traditions, is sadly being lost, yet orthodox medicine has so much to learn from it.
>
> (BBC news 2006)

He also urged countries across the globe to improve the health of their populations through a more 'integrated' approach to health care. What he failed to mention is the fact that integrating disproven therapies into our clinical routine, as proponents of 'integrated medicine' demonstrably do, will not render medicine better or more compassionate, but worse and less evidence-based (Ernst 2013).

While Charles' advocacy for CAM is well-known, his promotion of non-validated diagnostic methods, like those in abundant use in CAM, is not generally appreciated. Such methods run an unacceptably high risk of producing false positive or false negative diagnoses. As discussed previously, the former would be a diagnosis that the patient is, in fact, not suffering from. The latter would be missing an illness that might even kill the patient.

Charles published his book 'Harmony' which covers (amongst many other topics) the subject of alternative diagnostic methods (H.R.H. Prince of Wales et al. 2010). Here is a sample of what he wrote:

> I have also learnt from leading experts how we can understand a great deal about the causes of ill health through more traditional methods of diagnosis – for example, through examination of the iris, ears, tongue, feet and pulse, very much the basis of the Indian Ayurvedic system. This is not to say that modern diagnostic techniques do not have a role, but let us not forget what we can gain by using the knowledge and wisdom accumulated over

thousands of years by pioneers who did not have access to today's technology. In fact, an over-reliance can often mean that the subtle signs of imbalance revealed by the examination of the eyes, pulse and tongue are totally missed. Including the fruits of such knowledge, gleaned over 8 000 years of studying the relationship of the human body to the rest of Nature and to the Universe, can but only provide an extra, valuable resource to doctors as they seek to make a full diagnosis. Why persist in denying the immense value of such accumulated wisdom when it can tell us so much about the whole person – mind, body and spirit? Employing the best of the ancient and modern in a truly integrated way is another example of harmony and balance at work.

Charles is talking here about iridology, amongst other CAM methods—a diagnostic technique we discussed previously. Given that the evidence for iridology and other alternative diagnostic techniques is either negative or absent, why does the heir to the throne advocate using them? Does he not know that he has considerable influence and endangers the health of those who believe him? Why does he call this nonsense valuable? The answer probably is that he does not know better.

There is nothing intrinsically wrong with Charles' ignorance, of course. He is no physician and does not need to know such things. But, if he is ignorant about certain technicalities, he should not publish falsehoods like those cited above. He should, rather, recruit and listen to the expertise of people who do know about such matters. His failure to do so is objectionable not least on ethical grounds.

The antibiotics crisis is one of the urgent issues of our time and great attention should be paid to any possible solution. In 2016, Charles addressed a summit of experts on the issue of antibiotic over-use. The meeting took place at one of our most prestigious institutions of science worldwide, the Royal Society in London.

Charles, whose weakness for CAM is nowhere more apparent than in his long-standing love affair with homeopathy, told his audience that he had long been worried about the over-use of antibiotics, stating:

> it was one of the reasons I converted my farming operation to an organic, or agro-ecological, system over 30 years ago, and why incidentally we have been successfully using homeopathic — yes, homeopathic — treatments for my cattle and sheep as part of a programme to reduce the use of antibiotics... I find it difficult to understand how we can continue to allow most of the antibiotics in farming, many of which are also used in human medicine, to be administered to healthy animals... Could we not devise more effective systems where we reserve antibiotics for treating animals where the use is fully justified by the seriousness of the illness?

> (H.R.H. Prince of Wales 2016)

Charles seems to have a few good points here with which few would disagree. Sadly, he spoils it all by not being able to resist his passion for homeopathy. We would judge this as follows:

- Yes, we have over-used antibiotics both in human and in veterinary medicine.
- Yes, this has now gone so far that it now endangers our health.
- Yes, it is a scandal that so little has happened in this respect, despite us knowing about the problem for many years.
- No, homeopathy is not the solution to any of these problems.

Charles's highly diluted homeopathic remedies are of course merely placebos—when used with human patients. What about animals as patients? A recent systematic review assessed the efficacy of homeopathy in cattle, pigs and poultry (Doehring and Sundrum 2016). Here is a summary:

> Only peer-reviewed publications dealing with homeopathic remedies that could possibly replace or prevent the use of antibiotics in the case of infective diseases or growth promotion in livestock were included. Search results revealed a total number of 52 trials performed within 48 publications fulfilling the predefined criteria. Twenty-eight trials were in favour of homeopathy, with 26 trials showing a significantly higher efficacy in comparison to a control group, whereas 22 showed no medicinal effect. Cure rates for the treatments with antibiotics, homeopathy or placebo varied to a high degree, while the remedy used did not seem to make a big difference. No study had been repeated under comparable conditions. Consequently, the use of homeopathy cannot claim to have sufficient prognostic validity where efficacy is concerned. When striving for high therapeutic success in treatment, the potential of homeopathy in replacing or reducing antibiotics can only be validated if evidence of efficacy is confirmed by randomised controlled trials under modified conditions.

Perhaps the Royal Society could ask Charles for the evidence to support his claim? Because, given their motto (Nullius in verba), they cannot possibly take his word for it—that would hardly be ethical.

6   A 'Trustworthiness Index' for CAM?

The fact that researchers can be dishonest or wilfully misleading is neither new nor confined to CAM. Science has had its steady stream of scandals which are much more than just regrettable. They undermine much of what science stands for. In medicine, fraud and other forms of misconduct of scientists can even endanger the health of patients.

On this background, it would be good to have a simple measure to give us some indication about the trustworthiness of scientists, particularly clinical scientists, some type of a 'Trustworthiness Index' (TI).

Clinical science is often about testing the efficacy of treatments, and it is the scientist who does this type of research on whom we here intend to focus. Occasionally, clinical trials will generate negative results such as "the experimental treatment was not effective" (actually, 'negative' is not the right term, as it is clearly positive to know that a given therapy does not work). If this never happens with the work of a researcher, our alarm bells should start ringing, and we might begin to ask ourselves, how trustworthy is this scientist?

Yet, in real life, the alarm bells rarely do ring mostly because, at any one point in time, one single person tends to see only one particular paper of the individual in question—and one result tells us nothing about the question of whether this scientist produces more than his fair share of positive findings. What is needed is a measure that captures the totality of a researcher's output. If, for instance, we calculated the percentage of a researcher's papers arriving at positive conclusions and divided this by the percentage of his papers drawing negative conclusions, we might have a useful measure.

Consider, for instance, the case of a clinical researcher who has published a total of 100 original articles. If 50% had positive and 50% negative conclusions about the efficacy of the therapy tested, his TI would be 1. Depending on what area of clinical medicine this person is working in, a TI of 1 might be a figure that is just about acceptable in terms of the trustworthiness of the author. If the TI goes beyond 1, we might get concerned; if it reaches 4 or more, we should get worried. An example would be a researcher who has published 100 papers of which 80 are positive and 20 arrive at negative conclusions. His TI would consequently amount to 4. We might consider this figure to indicate a relatively low level of trustworthiness.

Of course, this is simplistic, and the TI fails to provide us with proof; it merely is an indicator that something might be amiss. Naturally, the cut-off point for any scientist's TI would depend on the area of clinical research that we are dealing with. The lower the plausibility and the higher the uncertainty associated with the efficacy of the experimental treatments, the lower the point where the TI might suggest something being suspect. An excellent example of an area plagued with implausibility and uncertainty is CAM. Here we would not expect a high percentage of rigorous tests to come out positive, and a TI of 0.5 might perhaps already be on the limit.

So how does the TI perform when we apply it to CAM researchers? One of us writes a blog (edzardernst.com) where he has evaluated the research of about a dozen prominent CAM researchers under the deliberately satirical title of 'The Alternative Medicine Hall of Fame'. Here we can see that, for all the individuals assessed in this way, the TI is excessively high; in some cases, the researchers have published *nothing but* trials with positive conclusion. This seems to indicate that the trustworthiness of a significant number of CAM researchers might be less than encouraging.

7 Conclusions

In the domain of CAM, truth is a frequent casualty. CAM practitioners, CAM researchers and CAM apologists—particularly when faced with critics who cite a lack of evidence of effectiveness—characteristically will not admit the manifest truth, but will instead strive doggedly to defend their beliefs, using more or less obvious untruths.

CAM defences against the truth are manifold. They include disingenuous attempts to deny the validity of inconvenient evidence. CAM proponents will often claim that mainstream research is just as flawed as CAM research—as if this would somehow justify sloppy CAM research. They may even make the astonishing 'postmodern' claim that objective truth cannot exist. If none of these defences work, CAM apologists have plenty of other tactics to call upon. They may try to reverse the burden of truth, illogically insisting that it is the responsibility of the critic to prove that CAM does not work. They will often claim that a 'paradigm shift' is imminent, which will allow the truth of CAM effectiveness to be accepted by scientists—a tactic that reveals a profound lack of understanding of the process by which scientific theories are established. Many CAM proponents assert that normal methods of medical research are not applicable for CAM, and want to go back to the dark ages of research by giving priority to 'personal experience'. They may introduce irrelevant claims to try to distract their opponents from the issue of ineffectiveness, such as the assertion that CAM is very safe compared with conventional medicine. And if none of the above works, then they can always fall back on making personal attacks against their critics, ranging from incompetence to corruption—seemingly unmoved by the fact that such allegations are rarely if ever evidence-based.

It is in the financial interests of CAM practitioners to recruit and retain patients, and in pursuit of this goal, and the income it brings, the truth is often further manipulated. Patients will often be told that they need treatment for non-existent conditions, or placed on 'health maintenance' programmes that are unnecessary, ineffective and in some cases potentially harmful. CAM practitioners will often employ falsehoods to deter patients from departing through a manifest lack of therapeutic effectiveness. Patients may be discouraged from returning to conventional medicine through the assertion that modern drugs have been the root of the patient's medical problem in the first place. Patients may be persuaded that, in CAM, symptoms must get worse before getting better; or that a CAM cure will inevitably take a very long time. Many CAM practitioners will cite holism to excuse therapeutic failure, attempting to plant in their patients' minds the false belief that other aspects of their health have been improved by the (ineffective) therapy. Finally, when pushed, most CAM practitioners will resort to justifying their practice on grounds of the placebo effect—which is essentially an admission of the truth that their magic therapy is ineffective.

Mangling of the truth is not restricted to the 'coal face' levels of CAM practice and research; proponents of CAM exist at all levels in society. This runs to the very top, where we have to endure Prince Charles publicly preaching in favour of homeopathy along with his other favourite anti-science superstitions. To knowledgeable rational observers, this spectacle is straightforwardly absurd; but because the heir to (and soon occupant of) the British throne is in a highly influential position, his pontifications do a disservice to the truth that is likely to have far reaching consequences.

The subversion of the truth entailed by CAM is an affront to medical ethics in whatever guise it occurs. From a purely utilitarian perspective, lying is only acceptable where more utility is likely to accrue from so doing than from truth-telling: in medicine and science, this is as a rule certainly not the case. The consequences of derogation of the truth in the domain of CAM are indubitably negative in terms of the effect of utility. By misleading patients, CAM at the very least wastes their time and resources, and at worst damages their health.

Beyond utilitarian considerations, subversion of the truth has additional ethical implications. Patient autonomy—a cornerstone of medical ethics—is undermined in line with the extent to which a practitioner uses falsehoods to induce the patient to submit to treatment.

The CAM practitioner who promotes untruths has either failed to enlighten themselves as to the facts—this being a central requirement of professional ethics— or has chosen to deliberately deceive patients. Either of these reasons for promulgating falsehoods amounts to a serious breach in terms of medical ethics.

According to almost all forms of ethical theory, the truth-violating nature of CAM renders it immoral in both theory and practice.

References

Ball P (2004) The memory of water. Nature. doi:10.1038/news041004-19

BBC news (2006) Doctors attack 'bogus' therapies. http://news.bbc.co.uk/1/hi/health/5007118.stm

Blease C, Colloca L, Kaptchuk TJ (2016) Are open-label placebos ethical? Informed consent and ethical equivocations. Bioethics 30(6):407–414. doi:10.1111/bioe.12245

Davenas E, Beauvaid F, Amara J et al (1988) Human basophil de-granulation triggered by very dilute antiserum against IgE. Nature 333(6176):816–818. doi:10.1038/333816a0

Doehring C, Sundrum A (2016) Efficacy of homeopathy in livestock according to peer-reviewed publications from 1981 to 2014. Vet Rec 179(24):628–641. doi:10.1136/vr.103779

Ernst E (2013) Prince Charles' vision of a "post-modern medicine" and my response to it. Edzard Ernst|MD, PhD, FMedSci, FSB, FRCP, FRCPEd. http://edzardernst.com/2013/01/prince-charles-vision-of-a-post-modern-medicine-and-my-response-to-it/. Accessed 21 Jun 2017

Genuis SJ, Schwalfenberg G, Siy AJ et al (2012) Toxic element contamination of natural health products and pharmaceutical preparations. PLoS ONE 7(11):e49676. doi:10.1371/journal.pone.0049676

Goldacre B (2013) Bad pharma: how medicine is broken, and how we can fix it. Fourth Estate, London

Grabia S, Ernst E (2003) Homeopathic aggravations: a systematic review of randomised, placebo-controlled clinical trials. Homeopathy 92(2):92–98

Howerter A, Hollenstein T, Boon H et al (2012) State-space grid analysis: applications for clinical whole systems complementary and alternative medicine research. Forsch Komplementarmed 19:30–35. doi:10.1159/000335187

H.R.H. Prince of Wales (2016) Speech to the global leaders conference on antimicrobial resistance. The prince of wales and the duchess of cornwall. https://www.princeofwales.gov.uk/media/speeches/hrh-the-prince-of-waless-speech-the-global-leaders-conference-antimicrobial. Accessed 21 Jun 2017

H.R.H. Prince of Wales, Juniper T, Skelly I (2010) Harmony: a new way of looking at our world. Blue Door, London

Kaptchuk TJ, Friedlander E, Kelley JM et al (2010) Placebos without deception: a randomized controlled trial in irritable bowel syndrome. PLoS ONE 5(12):e15591. doi:10.1371/journal.pone.0015591

Long L, Huntley A, Ernst E (2001) Which complementary and alternative therapies benefit which conditions? A survey of the opinions of 223 professional organizations. Complement Ther Med 9(3):178–185. doi: S0965-2299(01)90453-4 [pii]

Maddox J, Randi J, Stewart WW (1988) High-dilution experiments a delusion. Nature 334 (6180):287–290

Milgrom L, Chatfield K (2012) Is homeopathy really 'morally and ethically unacceptable'? a critique of pure scientism. Bioethics 26(9):501–503. doi:10.1111/j.1467-8519.2012.01948.x

Miller W, Crabtree B, Duffy M et al (2003) Research guidelines for assessing the impact of healing relationships in clinical medicine. Altern Ther Health Med 9(3):A80–A95

Oliver JE, Wood T (2014) Medical conspiracy theories and health behaviors in the United States. JAMA Intern Med 174(5):817–818. doi:10.1001/JAMAinternmed.2014.190

Paoli S, Freeman S (2010) Being against. Association Jacques Benveniste pour la Recherche. http://www.jacques-benveniste.org/Transcription-7-9.html. Accessed 04 Jun 2017

Paterson C, Baarts C, Launso L et al (2009) Evaluating complex health interventions: a critical analysis of the 'outcomes' concept. BMC Complement Altern Med 9:18. doi:10.1186/1472-6882-9-18

Rioux J (2012) A complex, nonlinear dynamic systems perspective on ayurveda and ayurvedic research. J Altern Complement Med 18(7):709–718. doi:10.1089/acm.2011.0569

Robinson M, Chorley M, Parry L (2015) Prince Charles demanded more homeopathy on the NHS. Mail Online

Russell B (1952) Is there a god? The campaign for philosophical freedom. http://www.cfpf.org.uk/articles/religion/br/br_god.html. Accessed 16 Jun 2017

Sokal A, Bricmont J (1999) Fashionable nonsense: postmodern intellectuals' abuse of science. St Martin's Press, USA

Verhoef M, Lewith G, Ritenbaugh C et al (2005) Complementary and alternative medicine whole systems research: beyond identification of inadequacies of the RCT. Complement Ther Med 13 (3):206–212. doi:10.1016/j.ctim.2005.05.001

Chapter 7
Exploitation

Exploitation can be defined as the use of another person or group for selfish purposes. Exploitative behavior in healthcare can lead to patients and others being harmed which clearly is unethical. In the context of this book, the following types of harm through exploitation are particularly relevant:

- Physical damage
- Mental distress
- Financial loss

Exploitation of vulnerable patients can impose physical damage. Many CAM therapies can inflict direct damage on a patient's body: examples include chiropractic manipulation resulting in paraplegia, acupuncture infecting patients with serious pathogens, and the toxicity of some oral remedies leading to severe liver damage. If a given CAM therapy is ineffective, any significant risk of physical damage renders its risk-benefit balance negative, and its use unethical. Moreover, where ineffective treatments are given to patients with serious diseases, physical damage occurs; in moral terms, such 'harm through omission' is just as unacceptable as direct harm.

Throughout this book, we have considered many instances of bodily damage that can occur using CAM. Accordingly, we shall not return to the theme of physical damage; rather, the remainder of this chapter will focus on harm through mental distress and harm through financial loss. The latter form of harm will receive the greatest attention, since it underpins the first two. In other words, if CAM purveyors ceased their efforts to financially exploit consumers, patients and society, the mental and physical harm inflicted by CAM would be largely eliminated.

© Springer International Publishing AG 2018   185
E. Ernst and K. Smith, *More Harm than Good?*
https://doi.org/10.1007/978-3-319-69941-7_7

1 Mental Distress

1.1 Made to Feel Guilty by CAM Practitioners

Some forms of CAM can be difficult or even impossible to adhere to. It is easy to see how attractive these types of CAM are for practitioners, as they provide an inbuilt get-out clause for the practitioner: "if only the patient had adhered to the treatment plan it would have worked". This leaves the practitioner off the hook for their exploitative behavior. Meanwhile, those who attempt to follow such treatments are likely to experience guilt because of their 'failure' to stick with the prescribed regimen.

An example is the Gerson® therapy, developed by Max Gerson in the 1930s, initially as a treatment for (his own) migraines, and subsequently as a treatment for cancer and other diseases. Gerson® therapy is presently advertised as:

> …a natural treatment that activates the body's extraordinary ability to heal itself through an organic, plant-based diet, raw juices, coffee enemas and natural supplements.
>
> With its whole-body approach to healing, the Gerson Therapy naturally reactivates your body's magnificent ability to heal itself – with no damaging side effects. This a powerful, natural treatment boosts the body's own immune system to heal cancer, arthritis, heart disease, allergies, and many other degenerative diseases.
>
> (Gerson Institute 2017)

However, there is no good evidence for any of these claims. The treatment consists of consuming copious quantities of juices and raw foodstuff as well as having coffee enemas. Particularly for patients suffering from cancer, this extreme diet is almost impossible to follow, yet strict adherence is said to be mandatory. When the cancer turns out to not be cured, patients are being told that they themselves are to blame: had they followed the regimen as instructed, this would not have happened. In the end, cancer patients not only die but also are made to feel guilty of their 'failure'.

CAM proponents can also induce feelings of guilt even amongst those who do not suffer from a disease. For example, the German 'Association of Catholic Doctors' (Bund Katholischer Ärzte) claim that homeopathic remedies can cure homosexuality. Of course, homosexuality is not a disease, but on their website, these doctors advise that (translation by EE):

> …the working group HOMEOPATHY of the Association notes homeopathic therapy options for homosexual tendencies…repertories contain special rubrics pointing to characteristic signs of homosexual behaviour, including sexual peculiarities such as anal intercourse.
>
> (Bund Katholischer Ärzte 2017)

In this context, they even state that "*homeopathy is not a straw to cling to…but a valuable instrument to help man in peril*". It goes without saying that these statements are complete and offensive nonsense. We would argue that they amount to an attempt to exploit gay men by inducing them to have their sexual orientation 'corrected' which will likely provoke strong feelings of guilt.

The inducement of guilt is a serious ethical issue. Utilitarian ethical theory is particularly relevant here: the avoidance or minimisation of negative mental states, such as feelings of guilt, is central to utilitarianism. Exploitative behaviours that tend to induce guilt, such as the sorts of CAM treatments described above, are therefore unethical in utilitarian terms.

1.2   False Hope

Hope is often essential for a patient's recovery from illness. False hope, on the other hand, is unlikely to be helpful; and when it is induced, serious ethical issues arise, as the patient is being lied to and their autonomy is thus being undermined. Moreover, when patients realise there they were given only false hope, and the CAM therapy has not worked, they are likely to experience a great deal of anguish which reduces their quality of life and arguably makes their situation worse.

Yet many promoters of CAM appear to have no qualms about giving false hope to vulnerable people. Such false hope is particularly objectionable when it is given to patients suffering from a life-threatening disease, such as cancer, and when it comes from 'high' or apparently authoritative places.

In 2004, Prince Charles publicly supported the above-mentioned Gerson® treatment, claiming that it is a hopeful treatment for cancer (BBC News 2004). Prof Baum, at the time one of the UK's most eminent oncologists, was invited by the British Medical Journal to respond to the Prince in an open letter. He wrote:

> ...Over the past 20 years I have treated thousands of patients with cancer and lost some dear friends and relatives to this dreaded disease...The power of my authority comes with knowledge built on 40 years of study and 25 years of active involvement in cancer research. Your power and authority rest on an accident of birth. I don't begrudge you that authority but I do beg you to exercise your power with extreme caution when advising patients with life-threatening diseases to embrace unproven therapies.

> (Baum 2004)

Despite the fact that they have attained their high positions merely through accidents of birth, monarchs undoubtedly have a good deal of influence over their 'subjects'. It is therefore inescapable that many cancer patients will have been given false hope by the utterances of Prince Charles. Accordingly, we consider his public support for unproven cancer treatments to be both foolish and immoral.

It is not only unelected monarchs who negatively influence the populace with quackery; some of our elected representatives in parliament are similarly inclined. For example, the Scotsman newspaper reported that David Tredinnick, the Tory MP for Bosworth, told his fellow MPs in 2016:

I was talking there to practitioners about what they are able to do for cancer patients, and there is actually a very long list of types of cancer that can be treated using traditional Chinese herbal medicine. One, cervical cancer, two, non-Hodkins lymphoma, three, HIV, four, colon cancer, five… six, breast cancer, seven, prostate cancer. And so the list goes on. I have in my constituency several constituents who I believe are alive today because they have used Chinese medicine. And the reason for that is what it does is it strengthens your system, and it strengthens the immune system, and it is very effective after cancer treatment. It deals with particular symptoms.

(Wheeler 2016)

It is obvious that none of these therapeutic claims are supported by good evidence. Tredinnick is, sadly, but one example; many other politicians, in the UK, US and beyond, are on record praising CAM. Such public statements from prominent elected representatives will inevitably provide false hope to many desperate patients.

Even medically qualified physicians are not immune to giving false hope through the promotion of quackery. Take for instance Dr. Elizabeth Thompson, a qualified doctor and leading UK medical homeopath. In the context of cancer in women, Thompson writes:

…Some people come when conventional treatments can no longer offer them anything to save their lives. This is a frightening time for them and although the homeopathic approach may not offer a cure at this late stage of their illness, it can often offer hope of a different kind. Sometimes it helps people to outlive the prognosis given to them by months or even years…

(Thompson 2017)

As we have discussed repeatedly, there is no good evidence that homeopathy has any health effects, let alone the ability to prolong the life of cancer patients. When physicians suggest otherwise, they inevitably give false hope to patients. Surveys have consistently shown that the public place more trust in medical doctors than they do in any other professionals, meaning that what doctors say tends to be believed and their advice is often closely followed. Thus, the false faith in CAM instilled by physicians such as Thompson amounts to a serious ethical violation.

In conclusion, exploitative behaviour by CAM advocates and practitioners tends to lead to mental distress. Exploitation can be for financial, career, or ideological purposes, where a CAM therapist wants to sell bogus treatments, or a CAM practitioner wants to be seen to be doing their job of treating patients (and thus continue earning a salary, or perhaps be promoted). Exploitation can also take the form of prominent individuals attempting to use their position or authority to influence people to use CAM, or to promote its use in public healthcare systems. In these cases, monetary or career gain need not be involved: different motivations lie behind such exploitative behaviour, including a desire to have one's own views writ large, and a wish to feel powerful. Regardless of their motives, those who cause mental distress by promoting CAM are, in our view, behaving in an ethically reprehensible manner.

2   Financial Loss

The sale of an ineffective treatment inevitably entails a monetary loss for the purchaser. Such financial exploitation is perpetrated across the entire spectrum of CAM providers, ranging from individual CAM practitioners operating alone, to clinics offering CAM treatments, to large chain stores or companies selling and manufacturing CAM medicines. Those suffering financial exploitation from the promotion of CAM range from individual citizens to society at large.

This section will focus on monetary loss occurring directly through the sale of CAM goods and services. It is worth noting in passing, however, that it is not just customers and patients who suffer financial exploitation in the world of CAM. As we discussed in Chapter 1, students can be innocent victims of college and university CAM programmes that induct them into a world of pseudoscience and futile therapeutic practice. Some of these individuals will financially benefit from their 'education', going on to work as CAM practitioners or setting up businesses selling quackery—in turn exploiting customers and patients, in ways we shall discuss below.

But other graduates from these programmes—perhaps those who are innately less gullible, or more ethically minded—will withdraw at some stage, when the scales drop from their eyes. Many of these unfortunate individuals thereby suffer a double financial hit: money spent on their studies (including course fees and living costs) has been sunk with no return on the investment, and their future earning potential may be damaged by dint of having to change career. And if their change of direction entails recommencing another education programme, yet more costs will be incurred.

2.1   *Financial Exploitation of Consumers*

According to the US National Institutes of Health, Americans spent in 2007 almost $34 billion out of their own pockets (i.e., not reimbursed by health insurance) on CAM, almost $12 billion of which was spent on an estimated 350 million visits to various CAM practitioners (NIH 2009). The remaining $22 billion was spent on "natural" products for self-care such as fish oils, plant extracts, glucosamine and chondroitin. On top of this, sales of vitamin and other nutritional supplements have been estimated to amount to a further $30 billion annually.

In many other countries, a similar situation prevails. A survey by the Fraser Institute, an independent, non-partisan Canadian public policy think-tank, found that Canadians spent $8.8 billion on CAM in 2016, up from $8 billion (inflation adjusted) in 2006 (Marketwired 2017). In the UK, it was estimated that £1.6 billion was spent on CAM in 2000 (Ernst and White 2000). Worldwide, the data that are available point to one conclusion: vast amounts of money are spent each year on CAM by consumers.

Throughout this book we have considered many CAM therapies, the majority of which are implausible and lack evidence of effectiveness. Whenever one of these therapies is sold to an individual, that person suffers a pecuniary loss. Therefore, it could be argued that virtually *all* practitioners of unproven modalities are engaged in financial exploitation.[1] However, this conclusion can be countered by the claim that it is the desires of individuals which drive the supply of CAM therapies; in other words, CAM practitioners are merely responding to commercial demand, as opposed to actively exploiting their customers.

Moreover, because the ethical principle of autonomy holds that competent individuals ought to be given maximum freedom to decide for themselves—even where they may make mistakes, such as purchasing an ineffective CAM remedy—sellers of CAM products and services could be viewed as providing a social benefit, in terms of enabling individual customers to exercise their autonomy. However, the assertion that a CAM seller is simply responding to demand (and thereby benefitting customers) is undermined to the extent that any of the following precepts fail to apply to the CAM service or merchandise proffered:

- Claims used to promote the product are accurate and truthful.
- Advertising is non-aggressive and does not take advantage of customer naivety or desperation.
- The product is 'merchantable'—i.e. it actually functions as it should.
- The sale will not harm third parties.

Below we shall consider several ways in which CAM purveyors routinely fail to meet the above precepts.

2.2 Exploitative Marketing of CAM to Healthy Consumers

When merchandises or services are advertised, it is often difficult to draw a diving line between [a] the provision of information intended simply to make the public aware of what is on offer, and [b] marketing that actively seeks to persuade individuals to make a purchase. Such a distinction may not matter: advertising of the latter type is generally accepted (if not wholeheartedly welcomed) by modern society—indeed it serves as a major economic driving force in consumer-based economies. However, a line does need to be drawn between these legitimate forms of advertising and promotional activity that could be described as unduly aggressive. In the context of CAM, unduly aggressive marketing usually involves exploitation of people's health-related fears, using dubious or false medico-scientific claims. Below we discuss two areas notorious for exploitation by sellers of CAM: anxiety about 'toxins', and distress about being overweight.

---

[1]CAM practitioners offering pro bono services would be an exception. But such altruism is rare.

'Detox' is a paradigm example of an area in which a great deal of exploitative CAM promotion occurs. As discussed in Chapter 4, detox treatments supposedly rid the body of toxins—yet the idea that people generally need 'detoxification' is unfounded, and claims that CAM treatments can remove toxins are not supported by good scientific evidence. But these facts do nothing to deter sellers of CAM from vigorously promoting a multitude of techniques for supposedly detoxifying our systems, including Ayurvedic medicines, colonic irrigation, special diets, ear-candles, foot-baths, homeopathic medicines, manual techniques, naturopathic medicines, and many other forms of quackery. Even Prince Charles launched his 'Duchy Originals Detox Tincture' a few years ago (Boseley 2009). It seems that virtually every CAM modality claims to be able to detox the system!

Detox purveyors aggressively market their merchandises and services by playing heavily on an anxiety that is commonplace amongst the public: namely, fear of bodily contamination. CAM proponents like to scare people into becoming customers with the spectre of the body harbouring nasty chemicals[2] from various sources, including industrial pollution, pesticide residues, pharmaceutical drugs, and junk foods. This is clearly exploitative, and hence unethical, because (as previously discussed) the actual dangers from these 'contaminants' are for most people grossly overblown; and in any case, there is no reliable evidence that any of the myriad CAM detox treatments proffered are able to do anything whatsoever to remove toxic chemicals from the body. (Of course, this last fact is unsurprising given the fundamental implausibility of most of these so-called detox methods.)

While many people are worried about 'toxins', it is likely that an even higher proportion of the populace is worried about being overweight. The world is experiencing an obesity epidemic, and CAM entrepreneurs have long jumped on the bandwagon of commercial opportunity provided by this malaise. Many slimming products and services, such as low-calorie food products and conventional weight-loss programmes, are valid enough, and their promotion is generally not exploitative. However, the market for weight loss is ripe for exploitation by quacks: regulations are lax or non-existent, and there are plenty of overweight people who are keen to believe anything they hear. It is, of course, difficult to lose weight, and people who want to slim are therefore often despairing, ready to try anything, and willing to pay unjustifiable amounts of money in the hope of a solution. The opportunities for financial exploitation are thus enormous. Consequently, a multitude of CAM weight loss treatments are being aggressively marketed to desperate and often gullible consumers.

The general premise that weight-loss can be induced by taking a pill or dietary supplement is a beguiling one. Of course, it would be much nicer to pay for a potion than to have to diet and exercise for weeks on end! That is, if these products were

---

[2]The term 'chemical' is frequently used a scare word by CAM proponents. But such usage demonstrates scientifically illiteracy: *all* matter is composed entirely of chemicals. Aside from forms of energy, *everything* in our environment—all solids, liquids, and gases—is made of chemicals.

effective, as claimed by their sellers. Sadly, however, the scientific facts simply do not support the bogus claims.

Prof. Ernst's research team conducted a programme of systematically assessing the efficacy and safety of alternative slimming aids. Their published analyses include the following treatments: African bush mango (Onakpoya et al. 2013), calcium supplements (Onakpoya et al. 2011d), chitosan (Pittler et al. 1999), chromium picolinate (Pittler et al. 2003), conjugated linoleic acid supplements (Onakpoya et al. 2012), garcinia extracts (Onakpoya et al. 2011b), green coffee (Onakpoya et al. 2011c), guar gum (Pittler and Ernst 2001), and *Phaseolus vulgaris* (Onakpoya et al. 2011a). In addition to these remedies and dietary supplements, acupuncture was also assessed, since acupuncturists like to claim that needling is not only effective in treating various diseases but is also able to suppress appetite (Ernst 1997). The results invariably showed that the outcomes were not convincingly positive: either there were too few data, or there were too many flaws in the studies, or—as in most of the cases—the weight reduction achieved was nil or too small to be clinically relevant.

Unfortunately, the message from these scientific studies is vastly outweighed by the aggressively advertised claims of quacks selling their pseudoscientific weight loss treatments. So, despite the overwhelmingly evidence to the contrary, consumers continue in their droves to believe that 'alternative' slimming aids are effective. What is more, consumers are also led to assume these products, which are frequently marketed as 'natural', are risk-free. This latter assumption is, sadly, false: apart from the harm done to the patient's bank account, many alternative slimming aids are associated with side-effects, including heart palpitations, vascular injury, vision loss, psychiatric events, and liver damage (Pittler et al. 2005; Nazeri et al. 2009; Kim et al. 2013). In some cases, these effects are serious, and can even include death.

2.3 Marketing Via the Internet

Prior to the mid-1990s, CAM sellers had limited scope for advertising their wares. But CAM purveyors now have a powerful tool to exploit the public relentlessly and effectively, namely the Internet. Even small commercial CAM firms can easily set up slick marketing websites—and link these to the similarly glossy websites of CAM associations/societies to appear more authoritative. And social media is being used ever increasingly to reach potential customers. The Internet provides a vast array of examples of this kind of exploitative promotion. Below we give a few examples from the morass of instances of Internet-facilitated exploitation by CAM purveyors.

A recent academic study examined the content of postings by chiropractors on the social media platform Twitter (Marcon et al. 2016). In the context of the efficacy and risks of chiropractic Spinal Manipulative Therapy (SMT), the researchers examined whether and to what extent: [a] debate occurs; and [b] critical information is disseminated. A sample of 1267 tweets was analysed, and the authors drew the following conclusions:

> In the abundance of tweets substantiating and promoting chiropractic and SMT (spinal manipulative therapy) as sound health practices and valuable business endeavors, the debates surrounding the efficacy and risks of SMT on Twitter are almost completely absent. Although there are some critical voices of SMT proving to be influential, issues persist regarding how widely this information is being disseminated.

Given that Twitter is accessible to everyone, large numbers of potential customers may be exposed to the kind of misinformation revealed by this study. Anyone pondering chiropractic care who follows such tweets is likely to be misled into imagining that SMT is safe and effective: yet, as we have discussed previously, the polar opposite is closer to the truth. By engaging in such pseudo-debates on social media, chiropractors are likely to influence their Twitter followers into purchasing chiropractic treatment—which presumably is a major motivation behind the posting of such tweets. In so doing, chiropractors are guilty of exploitative behaviour.

Unsurprisingly, in the context of the fast-moving and vast arena of the Internet, there are few published academic studies into the use of social media and websites for pro-CAM purposes. However, anyone can pursue the Internet to readily find many examples of exploitative CAM marketing. Websites featuring household names and stars tend to feature at the top of search engine results pages, and celebrities are particularly good at persuading their fans to part with their cash. For example, movie actress Gwyneth Paltrow had long recommended that women use steam baths ('Mugwort V-Steam') to clean out their vaginas (Pells 2015), when she decided to claim that putting a ball of jade in their vaginas is good for women. These jade balls can be purchased from her online business. On her website, she advertises this as follows:

> The strictly guarded secret of Chinese royalty in antiquity—queens and concubines used them to stay in shape for emperors—jade eggs harness the power of energy work, crystal healing, and a Kegel-like physical practice. Fans say regular use increases chi, orgasms, vaginal muscle tone, hormonal balance, and feminine energy in general. Shiva Rose has been practicing with them for about seven years, and raves about the results; we tried them, too, and were so convinced we put them into the goop shop. Jade eggs' power to cleanse and clear make them ideal for detox...

> (Paltrow and Rose 2017)

Browsing Paltrow's website, one finds no end of purchasable CAM products, advertised with the same sort of meaningless pseudoscientific gibberish that she uses to promote her jade eggs. This promotion of ineffective and potentially dangerous products is not merely nonsense; it amounts to exploitation of the public.

The number of websites promoting CAM products is vast and ever-growing (currently around 50,000,000). They tend to be very well presented, and give the impression of offering valid products and services. They typically encourage visitors to sign-up for emails and social media feeds, which serve to further beguile potential customers on a regular basis. Many sites link to 'professional body' websites for whichever creed their products hail from; of course, the latter web-pages provide only highly positive accounts of the modality in question. Pseudoscientific claims that are not backed up with robust evidence abound on these webpages; the authors of many of these websites bolster their claims by citing cherry-picked references from the academic literature—an impressive-looking touch that serves to further misleads visitors.

In fact, the audacity of those who exploit patients via the Internet seems to know no bounds. One website even used the research of Prof Ernst for advertising the use of magnetic bracelets against pain (Superior Magnetics 2017). Here is the text in question:

> Magnetic bracelets are a piece of jewelry, worn for the therapeutic benefits of the magnetic field. Magnetic bracelets has been used successfully by many people for pain relief of inflammatory conditions such as arthritis, tendinitis and bursitis…

> A randomized, placebo controlled trial with three parallel groups, came to the conclusion: Pain from osteoarthritis of the hip and knee decreases when wearing magnetic bracelets. It is uncertain whether this response is due to specific or non-specific (placebo) effects. Tim Harlow, general practitioner, Colin Greaves, research fellow, Adrian White, senior research fellow, Liz Brown, research assistant, Anna Hart, statistician, **Edzard Ernst, professor of complementary medicine**.

It is true that one of Ernst's studies was vaguely suggestive of a specific effect of magnetic bracelets over and above placebo. However, the entrepreneurs seem to have conveniently forgotten a couple of things that were expressed clearly in the original paper (Harlow et al. 2004):

- The study was published in the Christmas issue of the BMJ which specialises in publishing odd findings for their entertainment value.
- It was not possible to discern whether the symptom improvements were specific effects or merely a placebo effect.

Most importantly, this was just one trial, and one swallow does not make a summer! As pointed out repeatedly, we should always consider the totality of the reliable evidence. Being conscientious researchers, Ernst's team did exactly that and conducted a systematic review which concluded as follows:

> The evidence does not support the use of static magnets for pain relief, and therefore magnets cannot be recommended as an effective treatment. For osteoarthritis, the evidence is insufficient to exclude a clinically important benefit, which creates an opportunity for further investigation.

> (Pittler et al. 2007)

A more extensive analysis of the modern-era tools of exploitation deployed by CAM purveyors is beyond the scope of this book; instead, we encourage readers to take a little time to search for themselves around the Internet. Many examples conforming to the above descriptions and examples will be readily found within just a few clicks. We are confident that readers who look at these sites will agree with us when we say that the CAM entrepreneurs responsible for such exploitative mis-information are guilty of ethically reprehensible behaviour.

2.4   *High Street Sales of CAM Products*

The Internet is not the only location where quackery is peddled. Shops devoted to the sale of CAM products, and in some cases services, are commonplace on the high streets of towns and cities across the world. Stores selling TCM, complete with diagnostic and treatment advice, are one of the most widespread examples of this. These outlets exploit customers by selling ineffective and sometimes harmful remedies, and this is, of course, to be deprecated on ethical grounds.

But these stores do at least offer a degree of transparency, in the sense that consumers know that they are entering the domain of CAM before they walk through the doors. More pernicious, and therefore more ethically problematic, is the sale of CAM products under the same roof as bona fide healthcare products are retailed. In many countries, high street pharmacies are guilty of this, peddling unproven and disproven CAM products alongside conventional drugs. Often, major well-known pharmaceutical chains are involved; this inevitably sends out a 'stamp of approval' to consumers, many of whom assume that a 'big name' store would never stock bogus medical products.

The UK provides a salient example. The pharmacy chain 'Boots' is a huge business; with around 1500 shops across the country, it is the largest and most respected high street chemist in the UK, and its shelves have for many years stocked CAM medicines (Lewis 2006). The questionable products sold by Boots include herbal remedies, aromatherapy oils, food supplements that allegedly improve health, 'natural' slimming aids, and devices that supposedly reduce pain. Particularly disturbingly, Boots shops stock homeopathic medicines (and Bach Flower Remedies), which are, of course, devoid of plausibility or effectiveness. This situation is worsened by the fact that these products are stocked alongside conventional medicines, giving consumers the false impression that both classes of medicine are of equivalent standing. This is clearly exploitative and in our view unethical behaviour.

To make things even worse, qualified pharmacists within these stores are known to proffer uncritical advice to customers regarding these products. For example, in the case of a parent seeking advice in a Boots pharmacy on treating their 5 year old child, described as suffering from diarrhoea for three days, the pharmacist had no compunction about recommending the use of a homeopathic remedy (Colquhoun 2006). Many other examples of ethically questionable CAM vending by Boots exist, including the promotion of: B vitamins for "*vitality*"; 'natural Kaneka CoQ10' (a food supplement) for "*boosting energy levels*"; and Lactium (a milk derivative) to "*help you cope with the stresses and strains of everyday life*" (Colquhoun 2017).

This exploitative peddling of quackery by a respected pharmacy chain is ethically contemptible. Various individuals and newspapers have over the years published exposés of Boots' unethical CAM promoting behaviour, and have contacted the company to ask searching questions; all to no avail, as the sale of quackery by Boots continues apace. Given this intransigent determination by a large corporation

—and Boots is not alone here—to ignore basic ethical constraints and exploit consumers, the obvious question is: should the sale of CAM be more tightly regulated, or even outlawed? The counterargument to this is that consumer autonomy ought to trump concerns about exploitation. As we have emphasised throughout this book, autonomy is a cornerstone principle of ethics; so, the pro-autonomy position, in terms of allowing CAM products to be sold freely, appears prima facie attractive. However, it is questionable whether this position is as valid as it seems at first sight, as we discuss below.

2.5   The Ethics of Commerce

Respect for individual consumer autonomy is usually best served by a free marketplace, in which individuals are, within broad limits, able to choose and acquire the merchandises and services they view as being those best suited to satisfy their own perceived needs. Accordingly, most ethicists view market freedom as an important presumption. Following this argument, legal restrictions on selling (and advertising) are only justifiable in very exceptional cases, such as narcotic drugs, pornography, or weapons. Such products are produced by what some business ethicists have termed 'controversial industries'—an expression that suggests an ethically questionable status. In a later section, we shall return to the question of whether some CAM commercial activities and organisations ought also to be classed under the 'controversial industry' banner.

Aside from 'controversial' products, unfettered commerce is ethically desirable because the sale and advertising of merchandises and services maximises autonomy. It is also ethically supported from the utilitarian standpoint of maximising societal happiness. Moreover, the competition engendered by a free consumer market will tend to hold prices down and drive innovation—features which again benefit society.

These marketplace principles apply to all non-controversial goods. So, in the context of healthcare, a diversity of conventional (i.e. plausible and proven) products—such as indigestion tablets, sticking plasters, or blood pressure monitors—is a good thing for consumers. People have diverse needs, and prefer different products depending on the value they place upon product features, such as price, convenience, and dosing.

Given the desirability of a free market in conventional healthcare products, it could be argued that the presence of CAM in the healthcare market is also desirable. Certainly, most national regulators appear to agree with this, since in most jurisdictions controls over the marketing of CAM products are few and far between. Given that so many people want and are prepared to pay for CAM, as evidenced by the above consumer spending figures, there exists an apparent case for the ethical permissibility (or even desirability) of a free market in CAM merchandises and services.

On closer inspection, however, this presumption does not withstand scrutiny. A central pillar of the ethics of commerce concerns *merchantability*. This means that the product offered for sale must actually work as advertised. And the majority of CAM products evidently do not work: they are implausible and, in many cases, clinical trials have shown no evidence of effectiveness. On this argument, the selling of CAM products is unethical and amounts to exploitation, because consumers are misled as to the innate usefulness of the goods purchased (Macdonald and Gavura 2016).

One counterargument to this conclusion is that CAM purveyors are oblivious to the lack of ineffectiveness of their products: they are merely satisfying market demand. Either they simply do not care (they are simply selling a product, in which they have no interest), or they are deluded as to the reality of their products' in effectiveness. But this counterargument fails: as we have previously argued, in healthcare it is not sufficient merely to act in good faith—the stakes are simply too high. The onus is on the vendor of CAM products to properly understand what they are dealing with. Moreover, the principle of merchantability implies that reasonable efforts ought to be taken by the vendor to ensure that the product on sale actually works.

Another counterargument is that the notion of effectiveness is frequently too nebulous to apply to CAM, because often the consumer is placing a significance on the product that transcends mere medical effectiveness—such as a cultural, religious or spiritual value. By analogy, suppose a customer buys a small religious statuette on the grounds that the purchase will, according to the religion concerned, help increase the likelihood that the buyer's prayers will be answered. Objectively, this is based on a fantasy: obviously the statuette cannot perform this function. But it is hard to argue that the statuette's seller has acted unethically or exploitatively.

On this basis, the argument runs, it cannot be unethical to sell CAM, since customers will often be looking for benefits other than medical effectiveness. For instance, if a customer purchases a 'healing crystal' necklace, this might be because the item fits with their 'New Age' lifestyle and values; in which case, the effectiveness of the crystal (in terms of its ability to 'heal') may be irrelevant.

But this counterargument also fails. If benefits other than clinical effectiveness are the basis for a given CAM merchandise, this needs to be clearly communicated to the buyer. But the reality is that CAM sales are almost exclusively based on claims of effectiveness in the domain of healthcare: virtually all CAM advertising material does not promote products simply for their supposed cultural, religious or spiritual value.

Related to the concept of merchantability is the ethical principle of *general honesty* in commerce: specifically, that sellers ought to *refuse to profit from customer ignorance*. This is a fundamental rule, and appeals to common-sense ethical intuition, as well as being supported by more formal ethical systems. Yet this principle is clearly flouted in most CAM transactions. Purveyors of CAM products have (or should have) a good knowledge of their products' effectiveness, and this gives them an advantage over most of their customers. Given that the effectiveness of CAM treatments is mostly low or zero, it follows that most CAM sellers are exploiting their customers by profiting from their ignorance (Macdonald and Gavura 2016). This is, of course, ethically reprehensible behaviour.

In conclusion, the notion of a free market in CAM products may seem appealing at first sight, but it does not withstand scrutiny. By exploiting the ignorance of their customers, and selling them products that do not work as advertised, CAM purveyors violate well-established ethical principles of commerce. These violations are ubiquitous: indeed, it is almost impossible to trade in CAM without transgressing these principles, since most CAM products are intrinsically unfit for purpose.

2.6   Financial Exploitation of Vulnerable Patients

So far, we have focused on the exploitation of the public at large, as CAM purveyors try to convince otherwise healthy people that they need quack remedies to detox, diet, maintain health or keep their genitalia clean. Such exploitation is bad enough, but exploitation of vulnerable, severely-suffering patients is intolerable.

Yet such exploitation is commonplace. To witness this, one need look no further than the Internet. For example, the website of a substantial institution called 'Castle Treatments', which purports to have over 1000 positive testimonials, is addressed at some of the most vulnerable and desperate (Castle Treatments 2017):

> We are the U.K.'s leading experts in advanced treatments to help clients to stop drinking, stop cocaine use and stop drug use. Over the last 12 years we have helped over 9,000 private clients stop using: alcohol, cocaine, crack, nicotine, heroin, opiates, cannabis, spice, legal highs and other medications…
>
> When compared to any other method there is no doubt our treatments produce the best results. Over the last 12 years we have helped over 9,000 clients the stop drinking, stop cocaine use or stop using drugs with excellent results…
>
> Our treatment method uses specific phase signals (frequency) to help:
>
> • neutralise any substance and reduce physical dependency
> • improve and restore physical & mental health
>
> The body (muscle, tissue, bones, cells etc.) radiate imbalances including disease, physical, emotional and psychological conditions which have their own unique frequencies that respond to various 'beneficial input frequencies' (Hz) or 'electroceuticals' which can help to improve physical and mental health hence why our clients feel so much better during/after treatment…

The idea that 'phase signals' or 'electroceuticals' have therapeutic powers is scientifically untenable; in fact, it is sheer pseudoscience. Unsurprisingly enough, there is no good evidence that these treatments are effective for drug or alcohol dependency (or anything else for that matter). Even thousands of testimonials do not amount to evidence: the plural of anecdote is anecdotes, not evidence! Claiming otherwise is, in our view, highly irresponsible. If we consider the hefty fees Castle Treatments charge (Alcohol Support: Detox 1: £2655.00, Detox 2: £3245.00, Detox

3: £3835.00) we feel justified to characterise this as serious financial exploitation of vulnerable patients.

The most deplorable exploitation must be that of desperate cancer patients. And yet hundreds, if not thousands of different cancer treatments are promoted by CAM advocates. We have already considered Gerson® therapy above, and elsewhere in this book we have mentioned various quack cancer 'cures'. Some CAM proponents even claim that homeopathy can cure cancer, despite homeopathic remedies being devoid of any active ingredients. For example, in article posted by 'The Homeopathic College', a sizeable outfit based in North Carolina, the following statement is included:

Laboratory studies in vitro and in vivo show that homeopathic drugs, in addition to having the capacity to reduce the size of tumors and to induce apoptosis, can induce protective and restorative effects. Additionally homeopathic treatment has shown effects when used as a complementary therapy for the effects of conventional cancer treatment. This confirms observations from our own clinical experience as well as that of others that when suitable remedies are selected according to individual indications as well as according to pathology and to cell-line indications and administered in the appropriate doses according to the standard principles of homeopathic posology, homeopathic treatment of cancer can be a highly effective therapy for all kinds of cancers and leukemia as well as for the harmful side effects of conventional treatment.

(Mueller 2012)

It is almost needless to stress that there is no good evidence to suggest that homeopathy is a cure for cancer (or any other condition). But there are thousands of Internet sites claiming otherwise, offering homeopathy and all sorts of other CAM products as 'alternative cancer treatments'—for good money, of course. In our view, this amounts to scandalous exploitation, particularly where these 'cures' cannot plausibly have any effect on the natural history of cancer.

The organisation 'CAM-cancer' aims to provide evidence-based information on 'alternative' treatments for cancer (CAM-Cancer 2017). Their website contains reports on a wide range of CAM cancer treatments, and their conclusions are mostly negative. Here are the key passages concerning some herbal cancer remedies:

Aloe vera: ...*studies are too preliminary to tell whether it is effective.*

Artemisia annua: ...*there is no evidence from clinical trials...*

Black cohosh: ...*In all but one trial black cohosh extracts were not superior to placebo.*

Boswellia: ...*No certain conclusions can be drawn...*

Cannabis: ...*The use of cannabinoids for anorexia-cachexia-syndrome in advanced cancer is not supported by the evidence...*

Carctol: ...*is not supported by evidence...*

Chinese herbal medicine for pancreatic cancer: ...*the potential benefit... is not strong enough to support their use...*

Curcumin: *There is currently insufficient documentation to support the effectiveness and efficacy of curcumin for cancer...*

Echinacea: *...there is currently insufficient evidence to support or refute the claims... in relation to cancer management.*

Essiac: *There is no evidence from clinical trials to indicate that it is effective...*

Garlic: *Only a few clinical trials exist and their results are inconclusive.*

Green tea: *...the findings... are still inconclusive.*

Milk vetch: *Poor design and low quality... prohibit any definite conclusions.*

Mistletoe: *...the evidence to support these claims is weak.*

Noni: *...evidence on the proposed benefits in cancer patients is lacking...*

PC-Spes: *...contamination issues render these results meaningless. An improved PC-Spes2 preparation was evaluated in an uncontrolled study which did not confirm the encouraging results...*

St John's wort: *...there are no clinical studies to show that St. John's wort would change the natural history of any type of cancer...*

Ukrain: *...several limitations in the studies prevent any conclusion.*

So far, CAM-Cancer have not identified a single herbal cancer treatment that positively affects the natural history of any form of cancer. Selling such 'cures' to desperate patients, as thousands of entrepreneurs do, is unethical exploitation.

2.7   Financial Exploitation of Society

Financial exploitation does not only concern those consumers and patients who buy bogus CAM products or services, it concerns us all. Health insurances often pay for CAM, as do socialised healthcare systems such as the NHS in the UK; as citizens, therefore, we are all being potentially exploited. In 2007, one of us published an analysis of German health insurance companies' policies regarding bogus treatments (Ernst 2007). For this purpose, three popular CAM modalities were selected: Bach flower remedies, applied kinesiology (AK), and Schüssler salts.

We discussed Bach flower therapy previously, and noted that the ultra-dilute medicines it employs are as hopeless as all other forms of homeopathy; AK involves testing 'muscle tension' in order to diagnose diseases (including many diseases that have nothing to do with muscles, such as asthma or cancer), which are then treated with a range of physical manipulation approaches and other implausible therapies; and 'Schüssler salts' comprise a mixture of homoeopathically diluted minerals which allegedly "activate self-healing powers" (Deutsche Homöopathie-Union 2017). All three of these CAM modalities are entirely implausible and are not supported by good evidence of effectiveness.

What emerged from Ernst's evaluation was shocking: of the 13 insurance companies considered, 9 paid for Bach flower remedies, 7 for kinesiology and 9 for Schuessler salts. This is ethically problematic, for two major reasons. Firstly, insurance premiums must bear the cost of these useless therapies, therefore the companies and individuals who pay for health insurance are being financially exploited. Secondly, when large organisations such as healthcare insurance companies appear to support CAM, this implies a 'stamp of approval', which can only help boost the status of these quack therapies in the eye of the public. One expects large respectable companies to act with due diligence, and for insurance companies specifically to evaluate evidence to determine their practice. When they fail to do so, as manifestly is the case, only CAM purveyors benefit: everyone else (including the insurance companies) is exploited. This shows that it is not only individuals who are susceptible to pro-CAM propaganda; large organisations can also be swayed by the persuasive influence of CAM ideologues and quackery sellers, either directly or through the demands of a misinformed customer base.

The situation in most other countries is not much better. A review from the US concluded that, although the effect on overall insurance expenditures was modest, the number of people using CAM insurance benefits was substantial (Lafferty et al. 2006). And a recent US assessment suggested that, while the rates of health insurance reimbursement for CAM providers are somewhat lower than that for primary care physicians, the level is considerable (Whedon et al. 2017).

The *Wall Street Journal* reported that over 80% of the money that Medicare paid to US chiropractors in 2013 went for medically unnecessary procedures, with the federal insurance program for senior citizens spending roughly $359 million on unnecessary chiropractic care that year, according to a review by the Department of Health and Human Services' Office of Inspector General (OIG) (Evans 2016). The OIG report was based on a random sample of Medicare spending for 105 chiropractic services in 2013. It included bills submitted to CMS through June 2014. Medicare audit contractors reviewed medical records for patients to determine whether treatment was medically necessary. The OIG called on the Centers for Medicare and Medicaid Services (CMS) to tighten oversight of the payments, noting its analysis was one of several in recent years to find questionable Medicare spending on chiropractic care. "Unless CMS implements strong controls, it is likely to continue to make improper payments to chiropractors," the OIG said.

The above cases are just examples of a common pattern of insurance companies squandering premium payers' money on anomalous forms of healthcare. In countries with socialised forms of healthcare provision, such as the UK, similar issues exist. For example, Freedom of Information requests in 2010 revealed that the NHS spends about £4 million annually on homeopathic remedies alone (Triggle 2010; Poling 2010). It even funds clinics and hospitals that specialise in CAM provision,[3] such as a homeopathic hospital in Glasgow—to which we shall return later in this chapter.

The UK *Telegraph* newspaper reported that "*homeopathic medicines will escape an NHS prescribing ban even though the Chief Medical Officer Dame Sally Davies has dismissed the treatments as 'rubbish' and a waste of taxpayers' money*" (Knapton 2017). Sandra Gidley, chair of the Royal Pharmaceutical Society, said:

---

[3]These establishments often use the terms 'integrative medicine' in their title; integrative being a weasel-word designed to (falsely) deflect suspicion of quackery and send a reassuring message that conventional medicine is also present.

"*We are surprised that homeopathy, which has no scientific evidence of effective-ness, is not on the list for review. We are in agreement with NHS England that products with low or no clinical evidence of effectiveness should be reviewed with urgency.*"

Julie Wood, Chief Executive of NHS Clinical Commissioners, the body which was asked to review which medications should no longer be prescribed for NHS England, said: "*Clinical commissioners have always had to make difficult choices about prioritising how they spend their budget on services, but the finance and demand challenges we face at the moment are unprecedented. Clinical Commissioning Groups have been looking at their medicines spend, and many are already implementing policies to reduce spending on those prescribeable items that have little or no clinical value for patients, and are therefore not an effective use of the NHS pound.*" Yet the NHS Clinical Commissioners could not offer an expla-nation as to why homeopathic medicines had escaped the cut.

The exploitation of socialised healthcare is certainly not confined to the UK. CBC news (Canada) reported in 2017 that, more than a decade ago, the Manitoba Chiropractic Health Care Commission had been tasked to review the cost effec-tiveness of chiropractic services. It therefore prepared a report in 2004 for the Manitoba province and the Manitoba Chiropractors Association. The report makes 37 recommendations, including:

- The province should restrict its funding of chiropractic therapy to treatment of acute lower back pain and minimal coverage of the treatment of neck pain.
- The report called the literature around the efficacy of chiropractic care for neck pain "ambiguous or at best weakly supportive" and noted such treatment carried a "not insignificant safety risk."
- Chiropractic treatment should not be funded for anyone under 18, as "*the literature does not unequivocally justify*" the "*efficacy or safety*" of such treatment.

(Nicholson 2017)

The report also challenged claims that chiropractic treatments can be used to address a wide variety of medical conditions. It stated that there was not enough evidence to conclude chiropractic treatments are effective in treating muscle ten-sion, migraines, HIV, carpal tunnel syndrome, gastrointestinal problems, infertility or cancer, or as a preventive care treatment. It also said there was not enough evidence to conclude chiropractic treatments are effective for children. The report urged Manitoba Health to establish a monitoring system to keep a closer eye on "*the advertising practices of the Manitoba Chiropractors Association and its members to ensure claims regarding treatments are restricted to those for which proof of efficacy and safety exist.*" It suggested the government should have reg-ulatory powers over chiropractic ads.

Despite its obvious importance, the report has been kept secret, and Manitoba chiropractors have been able to exploit the public by advertising their services for a wide range of conditions including Alzheimer's, autism and paediatric conditions.

Individuals and employers who pay for healthcare insurance at least have the choice of changing provider; taxpayers funding their nation's healthcare system cannot do so. Moreover, because the respectability and influence of publicly funded healthcare institutions is substantially greater than that of insurance companies, the undeserved approval of CAM implied by its public funding will inevitably be perceived by laypeople as being more valid than implicit approval from the private insurance sector. This contributes towards undermining scientific medicine, which will in turn result in intangible but real negative effects for healthcare and public health.

2.8   CAM as a 'Controversial Industry'

Social science and business ethicists have applied the term 'controversial' to certain products and the industries and that supply them. The precise criteria for applying this term have varied over time (Reast et al. 2013), but in general the designation is used to describe merchandises, services or commercial organisations that are either dangerous or, to use a classic definition: *"for reasons of delicacy, decency, morality, or even fear, elicit reactions of distaste, disgust, offence or outrage when mentioned or when openly presented"* (Wilson and West 1981). Examples of controversial industries explored in the literature include armaments, gambling, pornography, or tobacco. Should CAM belong on the list of controversial industries?

Aside from issues of taste and perceptions of decency, a fundamental ethical criterion for designating an industry as 'controversial' is the extent to which their products may exploit by causing harm. All of the foregoing examples can certainly harm the customer; and so can CAM, as we have documented extensively in this book. However, in terms of the ethical permissibility (or otherwise) of allowing members of the public to purchase the wares of such industries, a tension exists between the principles of [a] individual autonomy versus [b] prevention of harm to the customer.

We shall not discuss this libertarianism-paternalism debate further, because it is not the key issue. In fully liberal markets, in which customer autonomy trumps consumer protection, as well as in highly regulated markets, these industries breach another important ethical principle, namely: *do no harm to third parties*. In ethical terms, such third-party harm is a defining feature of controversial industries.

Third-party harm is evident in the above non-CAM examples: weapon sales to the public can lead to extra deaths, accidental and criminal; gambling can cause addicts to squander domestic resources and impoverish their families; pornography may psychologically harm young children who view it; and tobacco smoke damages the health of smokers and non-smokers. But does CAM inflict harm on third parties? If it does, it would provide a strong argument for classifying CAM as a controversial industry.

The anti-vaccination stance typically adopted by CAM proponents certainly inflicts third-part harm. As discussed in Chapter 1, prevention of disease usually requires close to 100% of the at-risk population being vaccinated in order that 'herd immunity' is attained, whereby the disease is maximally suppressed. Otherwise, the proportion (typically 5–10%) of vaccinated individuals who do not gain effective immunity are at risk of contracting the disease concerned. Many CAM practitioners promote their services as a form of 'preventive medicine' to maintain the health of families, and during these sessions advice will be offered by the practitioner: oftentimes this will include anti-vaccination advice. Whenever a CAM purveyor's anti-vaccination advice prevents a child from being immunised, that child's health is placed in jeopardy, as is the health of other children through diminution of herd immunity.

Many of these anti-vaccination CAM practitioners also sell 'alternative immunisation' treatments or products. For example, chiropractic manipulation may be offered; almost incredibly, many chiropractors believe that manipulating the spine of an infant can prevent infectious disease. Another example is 'homeoprophylaxis', in which homeopaths sell remedies known as 'nosodes'. These preparations are oral medicines comprising potentised bodily material such as pus, urine, blood, or faeces. There is no good evidence that any form of homeoprophylaxis is effective, which is unsurprising considering its gross implausibility as an alternative to vaccination (Ernst 2016).

Insofar as parents heed the advice from CAM purveyors, or buy their 'alternative vaccination' products, and refuse or delay vaccination for their children, third-part harm occurs. This fact alone provides a good reason to classify CAM as a 'controversial industry'.

A final way in which CAM commerce impacts on third parties is through damage inflicted on the natural environment by demand for animal components, frequently from endangered species, used to prepare traditional Chinese medicines and other quack remedies (Still 2003). For example, rhinos are being driven to extinction in large part due to the demand for their horns, the powder of which has supposedly medicinal properties. Rhino poaching had been relatively infrequent and stable for a decade, with only 13 animals being killed in 2007. But of late there has been an exponential rise in the number of rhinos being killed in South Africa (Save the Rhino 2017). The numbers killed jumped to 83 in 2008 and reached 1,175 in 2015. The TCM market is worth around $170 billion per annum and, as it creates employment for many, it seems to enjoy political protection in some parts of the world.

In sum, there are several strong reasons to consider CAM as an exploitative 'controversial industry'. Arguably, the hidden costs to society and a lack of accountability, together with misleading claims masquerading as legitimate medical treatments, render CAM particularly controversial in nature compared with other controversial industries. And while reasonable arguments can be mounted for removing some industries from the controversial list, the nature of CAM means that such arguments are unlikely to succeed. An example of a possibly redeemable industry might be pornography; it can be argued that—if its products can be keep out of sight of children, and if it does not exploit its employees—commerce in pornography is ethically acceptable because it provides a sought-after product that does not harm its adult customers. By contrast, because the products of CAM are almost entirely ineffective, and inflict direct and indirect harm, we see little reason why commercial CAM organisations should escape the 'controversial industry' moniker.

2.9   Strategic Attempts to Be Perceived as Morally Legitimate

Sociologists employ 'legitimacy theory' to interpret the strategic behaviour of organizations. In this context, 'moral legitimacy', as perceived by society, is essential for any organisation seeking resources to continue operating. To ensure the survival of their organisations, management in controversial industries actively seek moral support from society. They behave in calculated, persuasive ways designed to try to gain, maintain and repair moral legitimacy. Sociological research has documented such behaviour occurring in well-established controversial industries; and there is evidence that it also occurs in the less well-studied domain of CAM (Crawford 2016).

The UK Glasgow Homeopathic Hospital (GHH) provides a good example. After its founding by wealthy benefactors in the mid-19th, the hospital became part of the NHS in 1948. In the early decades of NHS status, the GHH was not held to account for its activities and no apparent challenge to its practices was mounted from society. Gradually, however, pressure mounted on the GGH, as regulatory threats emerged.

For example, in 2010 the UK House of Lords Science and Technology Select Committee (STSC) published a report concluding that individual experiential accounts of homeopathic effectiveness were due to nonspecific effects only, and recommended that using taxpayers' money to fund this practice on the NHS was unjustifiable (STSC 2010). Meanwhile, several skeptics were drawing public attention to the absurdity of homeopathy, and some activists were specifically questioning the public funding of the GHH.

Consequently, legitimising strategies emerged from GHH management to resist and repair negative perceptions about homeopathy as a legitimate medical modality, and thus secure continuity of funding from the public purse. These strategies included:

- Repackaging of classical experiential homeopathy into 'integrated care', to coexist along with other CAM and some biomedical modalities, with the GHH partially renamed, as the 'Glasgow Homeopathic Hospital: Centre for Integrative Care'.
- Striving to maintain inclusion of the GHH as part of the local hospital-based training programme for medical students.
- Moving to new premises amid a fanfare of awards for the inspiring medical building.
- Manipulative impression management through trenchant public attacks on critics; for example, claims that scientific studies used to discredit homeopathy had been cherry picked to advance the values of opponents (Brocklehurst 2014; BHA 2010).

It remains to be seen how this battle between opponents and advocates of homeopathy concludes; if the GHH is successful in repairing its perceived legitimacy to the extent that it will secure continued NHS funding, this will be a triumph of skilful strategy over sound ethics. The GHH is a paradigmatic example of a CAM 'controversial organisation': it lacks genuine moral legitimacy, and its modus operandi is the exploitation of taxpayers and patients.

3 Conclusions

When someone uses another person or group for selfish purposes, they are engaging in exploitative behavior. Sadly, such conduct is ubiquitous in the world of quackery. It indubitably amounts to a serious ethical issue. This is so from a number of ethical angles: CAM purveyors and proponents who exploit their customers or patients fall far short of the demands of ethically virtuous behavior, and from a utilitarian perspective, exploitative behavior in healthcare is highly problematic because it can lead to patients and others being harmed. The types of harm that can be inflicted by CAM include mental distress, physical damage and monetary loss.

In many cases, these three harms will coincide. For instance, consider a patient with cancer who has the misfortune to be persuaded to try Gerson® therapy in place of conventional treatment. When the bizarre cocktail of organic foods, 'natural' supplements and coffee enemas proves ineffective, as of course it must, the patient will be left in mental distress. This distress is due both to the disappointment that the promised cure failed to work, and because blame will fall on the patient for having failed to strictly adhere to the (impossible-to-follow) regimen. The patient will also have sustained physical damage, because during the time that the useless therapy was being attempted, the underlying cancer will have progressed. And monetary loss will also have occurred, because the therapy had to be paid for—most likely from the patient's own resources. (If the therapy is funded by health insurance or by a public healthcare system, monetary loss still occurs, albeit to those organizations and hence to society in general.) Moreover, the knowledge that physical and financial harm occurred is likely to exacerbate the emotional harm suffered by the patient.

The only possible ethical justification for selling ineffective CAM products is to appeal to the principle of autonomy. Free markets maximize autonomy, and therefore it can be argued that the sale of quackery ought not to be regulated, as it is up to consumers to decide what is best for them. While on first sight this argument seems attractive, several real-life factors call it into question. Firstly, there is an asymmetry of information between seller and purchaser: in most cases, consumers will not know what science says about the proffered CAM cure; the purchaser is therefore ripe for exploitation, if the seller cares to capitalize on their naivety. And the reality is that CAM purveyors certainly do take advantage of the ignorance of consumers—indeed they are virtually obliged to do so, as their sales would hardly be sustained if they told the truth about their products.

Those who seek to make money by ripping off and abusing their customers do so using a variety of clever but underhand marketing tools, ranging from pseudo-debates about CAM effectiveness on social media, slick Internet sites replete with pernicious lies about medical matters, and high street pharmacies stocking quack remedies in their 'pharmacy and health' sections alongside conventional medicines. The asymmetry of information that underlies and hence enables this sort of reprehensible selling undermines the claim that autonomy is best served by a free market in CAM.

Secondly, the selling of CAM violates established ethical mores of commerce. CAM fails the test of merchantability—the requirement that products and services must actually work as advertised. CAM also violates the principle that sellers ought to refuse to benefit from the customer ignorance. For customer autonomy to be enabled, it is essential that these principles of ethical commerce apply; the fact that this is not the norm in the domain of CAM is another strike against the notion that CAM sales ought not to be regulated.

Finally, aside from absolutist libertarianism, most systems of ethics maintain that some regulation is required to maximize autonomy. This is not as paradoxical as it sounds: if citizens are permitted to own lethal weapons and fire them wherever they choose, or drive their cars without any speed limits, third parties will suffer; and the harm to these innocent parties will not just be physical damage, it will also be a loss of the ability to exercise autonomy, due to injury (or death). Similarly, completely free access to addictive narcotic drugs such as heroin will damage third parties, such as the children of the resultant addicts; additionally, the users themselves suffer a loss of autonomy once addicted, because once they are under the sway of their addiction their scope for exercising free choice becomes radically curtailed.

Commercial organizations that sell weapons or addictive substances can be classed as 'controversial industries', and CAM purveyors arguably fall into this category, since they too sell products that exploit customers, damage third parties, and undermine genuine autonomy. It is notable that, faced with questions over the ethics of their behavior and existence, such organizations have been shown to respond strategically, to establish a perception of moral legitimacy amongst their customers or paymasters. CAM organizations that take such defensive actions are not trying to improve patient health or maximize customer autonomy: they are simply attempting to secure their position such that they can continue to profit from the exploitation of naive customers and desperate patients.

In conclusion, CAM sellers routinely exploit people. Given the nature of what is on sale, this exploitation is a virtually inescapable feature of the industry. To avoid the charge of failing to act in an ethically virtuous fashion, there is only one option for individuals in the domain of CAM to take: do not sell quackery. But the reality is that CAM purveyors choose to persist with their unethical trade. Accordingly, the conclusion is clear, at least in principle: government must legislate to regulate and restrict the sale of merchandises and services that do more harm than good.

References

ACSH (2016) Midwife, and homeopathy, faulted in home-birth death. American Council on Science and Health. www.acsh.org/news/2016/03/24/midwife-and-homeopathy-faulted-in-home-birth-death. Accessed 28 July 2017

Baum M (2004) An open letter to the Prince of Wales: with respect, your highness, you've got it wrong. Br Med J 329(7457):118. doi:10.1136/bmj.329.7457.118

BBC News (2004) Prince criticised over therapies. BBC Home. http://news.bbc.co.uk/1/hi/health/3876431.stm. Accessed 31 July 2017

British Homeopathic Association (2010) Response to the STSC on homeopathy. https://www.britishhomeopathic.org/wp-content/uploads/2013/08/ST-parts-1-6.pdf. Accessed 25 July 2017

Boseley S (2009) 'Make-believe and outright quackery' – expert's verdict on prince's detox potion. The Guardian. https://www.theguardian.com/uk/2009/mar/11/prince-charles-detox-tincture. Accessed 28 July 2017

Brocklehurst S (2014) Should the NHS pay for homeopathy? BBC News. http://www.bbc.co.uk/news/uk-scotland-19798824. Accessed 25 July 2017

Bund Katholischer Ärzte (2017) Homöopathie - kurz. http://www.bkae.org/index.php?id=946. Accessed 07 July 2017

CAM-Cancer (2017) The 'concerted action for complementary and alternative medicine assessment in the cancer field' project. www.cam-cancer.org. Accessed 19 July 2017

Castle Treatments (2017) About us. http://www.castletreatments.com/profile/. Accessed 20 July 2017

Colquhoun D (2006) Mis-education at boots the chemist. DC's improbable science. www.dcscience.net/2006/04/16/can-you-trust-boots/. Accessed 24 July 2017

Colquhoun D (2017) Search results for: boots. DC's improbable science. www.dcscience.net/page/1/?s=Boots. Accessed 24 July 2017

Crawford L (2016) Moral legitimacy: The struggle of homeopathy in the NHS. Bioethics 30 (2):85–95. doi:10.1111/bioe.12227

Deutsche Homöopathie-Union (2017) Dr. Schüssler Salts help regulate the balance of mineral salts within the cells of the body. http://www.schuesslersalts.com/en/how_do_they_work/how_do_they_work/index.html?p. Accessed 21 July 2017

Ernst E (1997) Acupuncture/acupressure for weight reduction? A systematic review. Wien Klin Wochenschr 109(2):60–62

Ernst E (2007) Health insurances pay for untested procedures. Falsely conceived "patient friendliness". MMW Fortschr Med 149(8):55–56

Ernst E (2016) Homeoprophylaxis, the homeopathic vaccine alternative, prevents disease through nosodes. Edzard Ernst|MD, PhD, FMedSci, FSB, FRCP, FRCPEd. http://edzardernst.com/2016/04/homeoprophylaxis-the-homeopathic-vaccine-alternative-prevents-disease-through-nosodes/ Accessed 19 July 2017

Ernst E, White A (2000) The BBC survey of complementary medicine use in the UK. Complement Ther Med 8(1):32–36. doi:10.1054/ctim.2000.0341

Evans M (2016) Medicare spent $359 million on unnecessary chiropractic care in 2013, audit finds. Wall Street J. Accessed 21 July 2017

Freckelton I (2012) Death by homeopathy: issues for civil, criminal and coronial law and for health service policy. J Law Med 19(3):454–478

Gerson Institute (2017) The Gerson therapy. https://gerson.org/gerpress/the-gerson-therapy/2017

Harlow T, Greaves C, White A et al (2004) Randomised controlled trial of magnetic bracelets for relieving pain in osteoarthritis of the hip and knee. Br Med J 329(7480):1450–1454

House of Lords Science and Technology Select Committee (2010) STSC evidence check 2: homeopathy. http://www.parliament.the-stationery-office.co.uk/pa/ld199900/ldselect/ldsctech/123/12301.htm. Accessed 25 July 2017

Kim EJ, Chen Y, Huang JQ et al (2013) Evidence-based toxicity evaluation and scheduling of Chinese herbal medicines. J Ethnopharmacol 146(1):40–61. doi:10.1016/j.jep.2012.12.027

Knapton S (2017) Homeopathic medicines to escape NHS prescribing ban. http://www.telegraph.co.uk/science/2017/03/28/homeopathic-medicines-escape-nhs-prescribing-ban/. Accessed 24 July 2017

Lafferty WE, Tyree PT, Bellas AS et al (2006) Insurance coverage and subsequent utilization of complementary and alternative medicine providers. Am J Manag Care 12(7):397–404. doi:3162 [pii]

Lewis A (2006) Boots the quack. The Quackometer Blog. www.quackometer.net/blog/2006/08/boots-quack.html. Accessed 24 July 2017

Macdonald C, Gavura S (2016) Alternative medicine and the ethics of commerce. Bioethics 30 (2):77–84. doi:10.1111/bioe.12226

Marcon RA, Klostermann P, Caulfield T (2016) Chiropractic and spinal manipulation therapy on twitter: case study examining the presence of critiques and debates. JMIR Public Health Surveill 2(2):e153. doi:10.2196/publichealth.5739

Marketwired (2017) Nearly eight in ten Canadians have used alternative medicines: survey. https://finance.yahoo.com/news/fraser-institute-news-release-nearly-090000762.html. Accessed 25 July 2017

Mueller M (2012) Is homeopathy an effective cancer treatment? The homeopathic college. http://thehomeopathiccollege.org/cancer-treatment/homeopathy-effective-cancer-treatment/. Accessed 20 July 2017

Nazeri A, Massumi A, Wilson JM et al (2009) Arrhythmogenicity of weight-loss supplements marketed on the Internet. Heart Rhythm 6(5):658–662. doi:10.1016/j.hrthm.2009.02.021

Nicholson K (2017) Publicly funded chiropractic care should have strict limits, leaked report says. CBC News. www.cbc.ca/news/canada/manitoba/publicly-funded-chiropractic-care-report-1.4076690. Accessed 24 July 2017

NIH (2009) Americans spent $33.9 billion out-of-pocket on complementary and alternative medicine. News Releases. https://www.nih.gov/news-events/news-releases/americans-spent-339-billion-out-pocket-complementary-alternative-medicine. Accessed 07 July 2017

Onakpoya I, Aldaas S, Terry R et al (2011a) The efficacy of *Phaseolus vulgaris* as a weight-loss supplement: a systematic review and meta-analysis of randomised clinical trials. Br J Nutr 106 (2):196–202

Onakpoya I, Hung SK, Perry R et al (2011b) The use of Garcinia extract (hydroxycitric acid) as a weight loss supplement: a systematic review and meta-analysis of randomised clinical trials. J Obes 2011:509038. doi:10.1155/2011/509038

Onakpoya I, Terry R, Ernst E (2011c) The use of green coffee extract as a weight loss supplement: a systematic review and meta-analysis of randomised clinical trials. Gastroenterol Res Pract 2011. Epub 31 Aug 2010. doi:10.1155/2011/382852

Onakpoya IJ, Perry R, Zhang J et al (2011d) Efficacy of calcium supplementation for management of overweight and obesity: systematic review of randomized clinical trials. Nutr Rev 69 (6):335–343. doi:10.1111/j.1753-4887.2011.00397.x

Onakpoya IJ, Posadzki PP, Watson LK et al (2012) The efficacy of long-term conjugated linoleic acid (CLA) supplementation on body composition in overweight and obese individuals: a systematic review and meta-analysis of randomized clinical trials. Eur J Nutr 51(2):127–134. doi:10.1007/s00394-011-0253-9

Onakpoya I, Davies L, Posadzki P et al (2013) The efficacy of *Irvingia gabonensis* supplementation in the management of overweight and obesity: a systematic review of randomized controlled trials. J Diet Suppl 10(1):29–38. doi:10.3109/19390211.2012.760508

Paltrow G, Rose S (2017) Jade eggs for your yoni. GOOP. http://goop.com/better-sex-jade-eggs-for-your-yoni/. Accessed 20 July 2017

Pells R (2015) Gwyneth Paltrow scorned for suggesting women steam-clean their vaginas. Independent. http://www.independent.co.uk/news/people/gwyneth-paltrow-scorned-for-suggesting-women-steam-clean-their-vaginas-10012004.html. Accessed 20 July 2017

Pittler MH, Ernst E (2001) Guar gum for body weight reduction: meta-analysis of randomized trials. Am J Med 110(9):724–730. doi:S0002934301007021 [pii]

Pittler MH, Abbot NC, Harkness EF et al (1999) Randomized, double-blind trial of chitosan for body weight reduction. Eur J Clin Nutr 53(5):379–381

Pittler MH, Stevinson C, Ernst E (2003) Chromium picolinate for reducing body weight: meta-analysis of randomized trials. Int J Obes Relat Metab Disord 27(4):522–529. doi:10.1038/sj.ijo.0802262

Pittler MH, Schmidt K, Ernst E (2005) Adverse events of herbal food supplements for body weight reduction: systematic review. Obes Rev 6(2):93–111. doi:OBR169 [pii]

Pittler MH, Brown EM, Ernst E (2007) Static magnets for reducing pain: systematic review and meta-analysis of randomized trials. CMAJ 177(7):736–742. doi:177/7/736 [pii]

Poling S (2010) NHS should pull homeopathic hospital cash. BBC News

Reast J, Maon F, Lindgreen A et al (2013) Legitimacy-seeking organizational strategies in controversial industries: a case study analysis and a bidimensional model. J Bus Ethics 118 (1):139–153. doi:10.1007/s10551-012-1571-4

Save the Rhino (2017) Poaching statistics. https://www.savetherhino.org/rhino_info/poaching_statistics. Accessed 19 July 2017

Still J (2003) Use of animal products in traditional Chinese medicine: environmental impact and health hazards. Complement Ther Med 11(2):118–122. doi:S0965229903000554 [pii]

Superior Magnetics (2017) New magnetic bracelet for pain relief. http://www.prweb.com/releases/magnetic/bracelets/prweb11902947.htm. Accessed 18 July 2017

Thompson E (2017) Cancer – female. British Homeopathic Association. https://www.britishhomeopathic.org/charity/how-we-can-help/articles/conditions/c/when-orthodox-medicine-has-nothing-more-to-offer/. Accessed 07 July 2017

Triggle N (2010) NHS money wasted on homeopathy. BBC News. http://news.bbc.co.uk/1/hi/health/8524926.stm. Accessed 18 Sept 2017

Whedon J, Tosteson TD, Kizhakkeveettil A et al (2017) Insurance reimbursement for complementary healthcare services. J Altern Complement Med 23(4):264–267

Wheeler R (2016) Tory MP: traditional Chinese medicine can treat cancer and HIV. The Scotsman. http://www.scotsman.com/news/tory-mp-traditional-chinese-medicine-can-treat-cancer-and-hiv-1-4312683. Accessed 07 July 2017

Wilson A, West C (1981) The marketing of unmentionables. Harv Bus Rev 59(1):91

Postscript

In this book, we have identified numerous and serious ethical problems with CAM. *The Moral Maze of Complementary and Alternative Medicine* is thus a biting criticism of CAM as it is practiced today. The reason why we criticise the 'today' is, of course, because we want to help create a better 'tomorrow'. So, what does the future hold for CAM? As scientists, we are not good at reading tea-leaves. Perhaps the better question therefore is: what should the future of CAM be, in view of the many ethical problems that we discovered?

In our view, the incessant violation of fundamental ethical principles is, in the long-run, not tolerable. Therefore, we would urge all involved—regulators, heath politicians, professional bodies, practitioners—to work towards bringing CAM in line with the currently accepted ethical standards of healthcare. In many instances, this might mean stopping CAM altogether. This will not be an attractive prospect to some, but the alternative would be to allow double standards in medical ethics. This, in our view, would not be in the best interest of consumers, patients or healthcare as a whole.

Our book will no doubt upset, anger or horrify many who are enthusiastic about CAM. They will disagree with and object to many of our arguments. But we do not mind constructive criticism at all. In fact, we welcome it in the hope that this eventually might lead to ethical progress. However, we fear that much of the criticism of our book—particularly that originating from the realm of CAM itself—will not be constructive but will take the form of ad hominem attacks. We predict that the more fanatical CAM proponents will claim that we are not qualified to judge in the domain of CAM, or that we are dishonest or corrupt.

It is therefore important, we feel, to point out that none of this is true. We are not paid by anyone to defame CAM. What we have written in this book are our sincere conclusions based on many years of research and experience. Our sole motivation in writing *The Moral Maze of Complementary and Alternative Medicine* is to generate ethical progress and contribute towards a better healthcare of tomorrow.

Glossary[1]

Acupuncture can be defined as the insertion of needles into the skin and underlying tissues at acupuncture points for therapeutic or preventative purposes. Traditional acupuncture is based on Taoist philosophy and has no grounding in science. Western acupuncturists believe that acupuncture is based on neurophysiological concepts. The effectiveness of acupuncture is still in dispute. Acupuncture frequently causes mild adverse effects and occasionally serious complications. For most conditions, the risk/benefit balance of acupuncture is negative

Anthroposophic medicine was developed by Rudolf Steiner about a century ago. It is based on mystical concepts that have no basis in fact. Numerous therapies are employed by anthroposophic doctors, few of which are supported by sound evidence. Depending on the nature of the therapy, anthroposophic treatments can cause minor or major adverse effects. The risk/benefit balance depends on the treatment in question but usually is not positive

Aromatherapy is the use of essential oils for medicinal purposes. There is a wide range of essential oils on offer. The term 'essential' here does not mean vital but is derived from essence. Aromatherapy is usually combined with gentle massage; less commonly it is applied via inhalation. Its effectiveness for specific diseases is not supported by good evidence, however, it is an agreeable and relaxing therapy. It can occasionally cause adverse effects such as allergic reactions. The risk/benefit balance might be positive if aromatherapy is used for relaxation and well-being

Ayurvedic medicine is a traditional Indian system of healthcare using oral medications, physical treatments, mind-body approaches and life-style measures.

---

[1]This glossary is aimed at providing short explanations about the CAM modalities mentioned in the text. The intention is to provide the essential details for understanding of the issues at stake. The main source for compiling the glossary was the *Oxford Handbook of Complementary Medicine* edited by one of the present authors (Ernst) in 2008 (Ernst E, Pittler MH, Wider B et al (2008) Oxford Handbook of Complementary Medicine. Oxford University Press, Oxford).

Some of the interventions involved might be effective for some conditions, but most are too under-researched to be sure. Many Ayurvedic medications contain toxic substances such as heavy metals and can therefore cause serious adverse effects. The risk/benefit balance depends on the exact modalities used

Bach flower remedies were developed about a century ago by Dr Edward Bach who had a background in homeopathy. They are based on his notion that all diseases are due to emotional imbalances which can be corrected with one of his 38 remedies. The remedies are too highly diluted to contain sufficient amounts of active ingredients. The evidence fails to show that they are efficacious. No direct risks have been reported. The risk/benefit balance is not positive

Bromelain is the name of a protein extract made from pineapples which is rich in enzymes. It is sold as a dietary supplement and claimed to work for a range of conditions. Some studies have suggested that it is efficacious for osteoarthritis; however, the evidence is not strong. Allergic reactions are possible. If used for osteoarthritis, its risk/benefit balance might be positive

Bryophyllum pinnatum This plant contains cardiac glycosides. It is native to Madagascar but has become a popular house plant in many countries. In some parts of the world, it is used as a traditional remedy for heart conditions. However, its effectiveness has not been well documented. In high doses, it can cause severe poisoning. Its risk/benefit balance is negative

Carctol is an herbal mixture developed by an Ayurvedic practitioner from India who has been promoting his remedy for many years as a treatment of a wide range of conditions. Today, Carctol is promoted mostly as a cancer cure. One 560-mg capsule of Carctol contains *Hemidesmus indicus* (roots): 20 mg/*Tribulas terrestris* (seeds): 20 mg/*Piper cubeba* Linn. (seeds): 120 mg/*Ammani vesicatoria* (plant): 20 mg/*Lepidium sativum* Linn. (seeds): 20 mg/*Blepharis edulis* (seeds): 200 mg/*Smilax china* Linn. (roots): 80 mg/*Rheumemodi wall* (roots): 20 mg. There is no evidence that it is efficacious. Believing the bogus claims could therefore cost cancer patients their lives. Its risk/benefit balance is negative

Chiropractic is a form of healthcare that was developed by D.D. Palmer and focusses on the relationship between the spine and the rest of the body. The hallmark therapy of chiropractors is spinal manipulation which, they believe, is necessary to adjust 'subluxations'. These are mystical entities which Palmer believed are the cause of virtually all disease. Consequently, he promoted chiropractic as a cure-all. Yet the effectiveness of chiropractic spinal manipulations is doubtful even for back and neck pain, which are their prime indications today. They cause transient adverse effects in about 50% of all patients; in addition, serious complications, including deaths have regularly been recorded. Its risk/benefit balance is usually not positive

Craniosacral therapy is a gentle form of manual therapy developed by W.G. Southerland and J.E. Upledger. They believed that rhythmic motions of the cerebrospinal fluid determine our health and can be influence by manual manipulations from the outside. The effectiveness of this approach is not supported by evidence. No adverse effects have been noted. Its risk/benefit balance is not positive

Cupping is a traditional therapy that has been used in several cultures. There are two different types of cupping. Dry cupping is a therapy where one or more vacuum cups are applied over the intact skin which usually is strong enough to cause bruising. In wet cupping, the skin is scratched prior to applying the vacuum which allows blood to be sucked into the cup. The mechanism of action might be that of a simple counter-irritation. Cupping is promoted for a wide range of conditions, usually with no or only scant evidence in support. Wet cupping can cause infections which can become serious, if untreated. The risk/benefit balance of wet cupping is negative, that of dry cupping is highly questionable

Eurythmy is an expressive movement and excercise originated about a century ago by Rudolf Steiner in collaboration with Marie von Sivers. It is used as a performance art, in education, and as an anthroposophic therapy. It is claimed to re-integration of body, soul, and spirit. However, it is neither biologically plausible nor supported by sound evidence. No specific adverse effects are known. Its risk/benefit balance is unknown

Evening primrose oil Evening primrose (Onagraceae family) is a plant native to North America. The oil made from its seed contains linoleic acids and is used therapeutically. It is commonly used as a supplement for a range of indications. However, for none of them is the evidence sufficient to justify its remarkable popularity. Adverse effects include headache and gastrointestinal symptoms. Its risk/benefit balance is negative

Gerson® therapy was developed by Max Gerson in the early 20th century. It includes a starvation diet of raw foodstuff and coffee enemas and is used mostly (but not exclusively) to treat cancer. There is no good evidence that it is effective for any condition and plausible evidence to suggest that it can cause severe harm. Therefore, its risks clearly outweigh its benefits

Herbal medicine or phytotherapy can be defined as the medicinal use of preparations that contain exclusively plant material. It is advisable to differentiate traditional herbalism from rational phytotherapy. Traditional herbalism is based on obsolete assumptions about the nature of disease and usually employs herbal mixtures that are tailor made for each patient. It is not supported by sufficient evidence for effectiveness. Rational phytotherapy is the use of a herbal extract based on evidence from clinical trials for specific condition. An example is St John's Wort for depression. The risk/benefit balance for evidence-based phytomedicine is positive

Homeopathy The basic principles of homeopathy were formulated by a German physician named Samuel Hahnemann in the early nineteenth century. Homeopathy is based on two central principles: the 'law of similars' and the 'law of infinitesimals'. The former principle holds that a substance able to cause a symptom in healthy subjects can also be used to cure that symptom. The latter principle holds that a therapeutic substance becomes more potent as it is diluted, provided that the process of dilution is accompanied by a special form of vigorous shaking ('succussion'). Hahnemann and his followers assembled a body of literature based on observations of the apparent effects of administration of a range of diluted substances on various subjects (including themselves). Its assumptions fly in the face of science and the trial evidence fails to support that it is more effective than placebo. Much harm is done not normally through the remedies but via the advice issued by homeopaths. Thus, the risk/benefit balance of homeopathy is negative

Homeoprophylaxis Homeopaths tend to advise their patients against immunisations and instead recommend homeopathic immunisations or 'homeoprophylaxis'. This normally entails the oral administration of homeopathic remedies, called nosodes, i.e. potentised remedies based on pathogenic material like bodily fluids or pus. There is no evidence that homeoprophylaxis is effective. After conventional immunisations, patients develop immunity against the infection in question which can be monitored by measuring the immune response to the intervention. No such effects can be observed after homeoprophylaxis. The risk/benefit balance is negative

Integrative care Integrative (or integrated) medicine has been defined as a 'comprehensive, primary care system that emphasizes wellness and healing of the whole person'. The aim of integrative medicine is to use evidence-based conventional and CAM side by side. The main problem with this approach is that most forms of CAM are not evidence-based. In practice, most centres of integrated medicine make ample use of unproven or disproven treatments

Intercessory prayer is the act of praying on behalf of others. In CAM, this is often used as a form of therapy where the ill person is prayed for by one or more people with the aim of improving his/her health. Intercessory prayer is not a plausible therapy and there is no good evidence that this approach is effective. The risk/benefit balance is not positive

Iridology Iridologists believe that discolourations on specific spots of the iris of a patient provide diagnostic clues as to the health of organs. They have maps where certain areas of the iris correspond to the organs of the body. Iridology is not in accordance with anatomical and physiological knowledge, and there is no good evidence that it can accurately diagnose anything; therefore, it is likely to lead to false-positive and false-negative diagnoses. The risk/benefit balance is negative

Kinesiology or applied kinesiology (AK) is a diagnostic technique developed by the US chiropractor George Goodheart Jr. It involves the practitioner testing the strength of the patient's muscle groups which allegedly provides information about the patient's health. The assumptions of AK are not plausible, and its validity has been tested repeatedly and was not confirmed. The risk/benefit balance of AK is therefore not positive

Laetrile also often called Vitamin B17 (although it is not a vitamin), is a partly man-made form of the natural substance amygdalin, found naturally in raw nuts and the pips of many fruits, particularly apricot, or kernels. It is converted to cyanide in the body and often claimed to be an effective cancer therapy. However, there is no evidence that it is effective and compelling evidence that it can be harmful. The risk/benefit balance is negative

Lens culinaris lectin Lentils are part of the staple diet in many regions of the world. They contain carbohydrate-binding proteins called lectins which are sometimes used for medicinal purposes. No serious adverse effects have been noted. The risk/benefit balance is unknown

Massage therapy Massage is a therapy where the soft tissues of whole body areas are manipulated, usually by hand. There are many types of massage therapies originating from different parts of the world. There is some evidence to suggest that some forms of massage are effective in treating some conditions. Massage therapy is not associated with serious adverse effects. Its benefits can outweigh its risks

Mindfulness-based stress reduction Mindfulness is the process of bringing one's attention to the internal and external experiences occurring in the present moment which can be developed through the practice of meditation. Mindfulness-based stress reduction is the use of mindfulness to assist people suffering from stress, pain or a range of other conditions. There is good evidence to suggest it is effective and safe. Therefore, the risk/benefit balance is positive

Mistletoe (*Viscum album*) therapy goes back to Rudolf Steiner and his anthroposophical philosophy. Steiner argued that mistletoe is a parasitic plant on host trees which it eventually kills, similar to a cancerous tumour killing a patient. Inspired by homeopathy's like cures like theory, he concluded that therefore, mistletoe must be a treatment for cancer. The evidence is conflicting but the best data fail to show that mistletoe preparations cure cancer or improve cancer patients' quality of life. Mistletoe is not free of adverse effects. The risk/benefit balance is not positive

Naturopathy is an eclectic system of healthcare which employs the forces of nature for stimulating the body's ability to heal itself. The modalities used include herbal extracts, manual therapies, heat and cold, water and electricity. Its effectiveness and safety depend on the exact mixture of modalities applied. The risk/benefit balance is in most cases not positive

Nutritional therapy Nutritional therapies use diets for medicinal purposes. In CAM, there is a plethora of diets which are being promoted as treatments for a wide range of conditions. They usually have in common that their plausibility, effectiveness and safety are unknown or uncertain. The risk/benefit balance of these nutritional therapies is often negative

Osteopathy is a manual therapy involving manipulation of the spine and other joints as well as mobilization of soft tissues. It was originated in the US by Andrew Still some 230 years ago. Today, US osteopaths have mostly become conventional physicians, while elsewhere they are CAM practitioners. Osteopathy is advocated for a wide range of condition, but only for back pain is the evidence encouraging. Osteopathy can cause adverse effects as well as serious complications. Its risk/benefit balance is usually not positive

Papain is an enzyme from papaya fruit that breaks down proteins. It is used medicinally for a wide range of conditions, both orally and topically. Its effectiveness is not supported by good evidence. No major adverse effects are on record. Its risk/benefit balance is uncertain

Phytotherapy is a synonym of herbal medicine. The term 'rational phytotherapy' is sometimes used to describe the use of evidence-based herbal treatments and differentiate them from the use of traditional herbal medicine which is usually individualised according to the characteristics of a single patient

Reflexology is the treatment employing manual pressure to specific areas of the body, usually the feet, which are claimed to correspond to internal organs with a view of generating positive health effects. The therapy lacks plausibility and its effectiveness is not supported by good evidence. There are no major risks. The risk/benefit balance fails to be positive

Reiki is a Japanese therapy where the therapist claims to channel life energy into the patient's body which is supposed to stimulate his self-healing abilities. The therapy lacks plausibility and its effectiveness is not supported by good evidence. There are no major risks. The risk/benefit balance fails to be positive

Schuessler salts were developed by the German homeopath W. H. Schuessler at the end of the 19th century. To this day, they are highly popular in Germany and are now beginning to find enthusiasts elsewhere. Schuessler believes that diseases were due to imbalances of the body's minerals. His remedies are highly diluted like homeopathic medicines. There is no evidence that Schuessler salts are more than placebos. Their risk/benefit balance is not positive

Shiatsu is a Japanese therapy akin to acupressure; the therapist uses his fingers to apply pressure to certain points of the body. The therapy lacks plausibility and its effectiveness is not supported by good evidence. There are few major risks. The risk/benefit balance fails to be positive

Sodium selenite is the sodium salt of selenium; it is a colourless, inorganic solid. Its effectiveness and safety are not known, and its risk/benefit balance is unknown

Spiritual healing is, like Reiki, a form of energy healing where the therapist claims to channel life energy into the patient's body which is said to stimulate his/her self-healing abilities. The therapy lacks plausibility and its effectiveness is not supported by good evidence. There are no major risks. The risk/benefit balance fails to be positive

Therapeutic touch is yet another form of energy healing where the therapist claims to channel life energy into the patient's body which is said to stimulate his/her self-healing abilities. It is particularly popular with US nurses. The therapy lacks plausibility and its effectiveness is not supported by good evidence. There are no major risks. The risk/benefit balance fails to be positive

Traditional Chinese medicine (TCM) is a diagnostic and therapeutic system based on the Taoist philosophy of Yin and Yang. It is used as an umbrella term for methods that emerged from China, including acupuncture, herbal medicine, tui-na (Chinese massage), tai chi and diet. Some of these treatments are biologically plausible, e.g. herbal medicine and diet. A few may be are effective and safe. The risk/benefit balance depends on the therapy in question but is generally negative

Ukrain is an alternative cancer drug based on two natural substances: alkaloids from the greater celandine and the compound thiotepa. Its effectiveness is highly questionable, although its adverse effects seem relatively minor. Its risk/benefit balance is not positive

Index

Printed in the United States
By Bookmasters